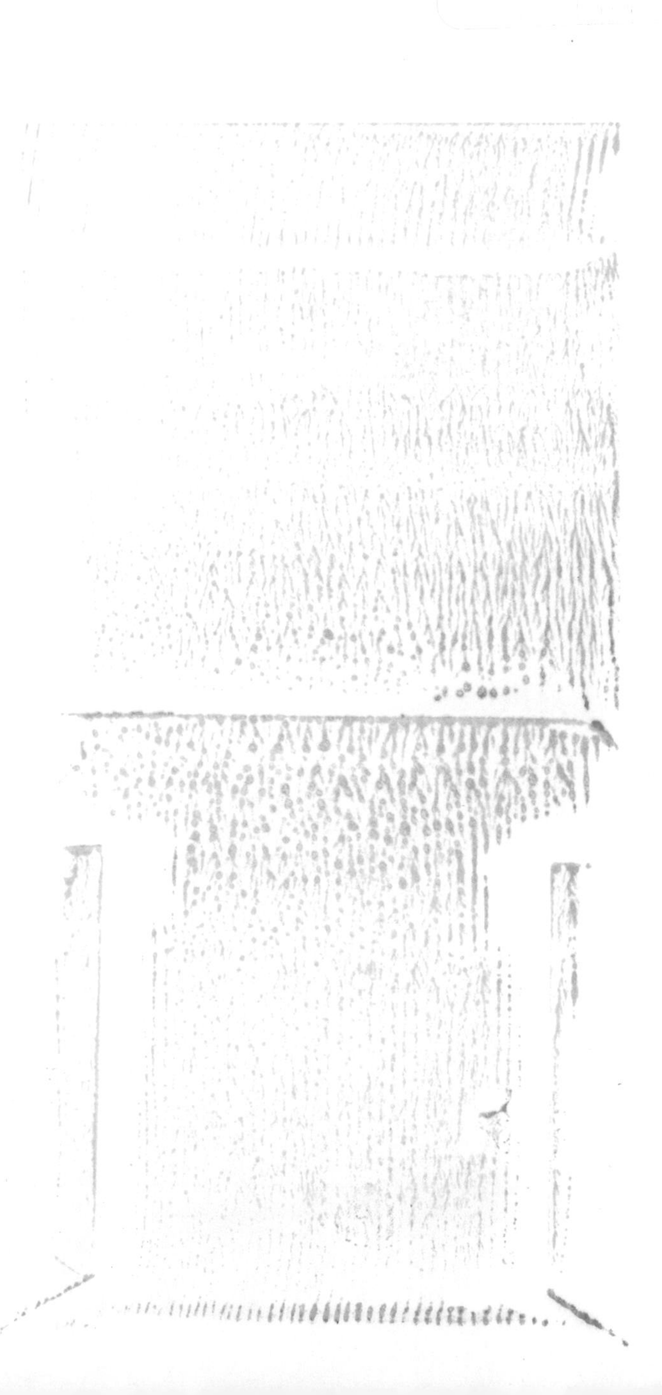

Introduction to Light

Introduction to Light

The Physics of Light, Vision, and Color

GARY WALDMAN

Professor of Physics
Florissant Valley Community College

Senior Staff Engineer
Emerson Electric Company

PRENTICE-HALL, INC., Englewood Cliffs, New Jersey 07632

Library of Congress Cataloging in Publication Data

Waldman, Gary.
Introduction to light.

Includes bibliographical references and index.
1. Light. I. Title.
QC355.2.W34 1983 535 82-16484
ISBN 0-13-486027-6

Editorial/production supervision
and interior design by Maria McKinnon
Cover: Image formed by scanning light from a 4-color Krypton laser.
Cover art by Floyd Rollefstad, Coherent Innovations, Inc.
Cover design by Lee Cohen
Manufacturing buyer: John Hall

Printed in the United States of America

10 9 8 7 6 5 4 3 2 1

ISBN 0-13-486027-6

PRENTICE-HALL INTERNATIONAL, INC., *London*
PRENTICE-HALL OF AUSTRALIA PTY. LIMITED, *Sydney*
EDITORA PRENTICE-HALL DO BRASIL, LTDA., *Rio de Janeiro*
PRENTICE-HALL CANADA INC., *Toronto*
PRENTICE-HALL OF INDIA PRIVATE LIMITED, *New Delhi*
PRENTICE-HALL OF JAPAN, INC., *Tokyo*
PRENTICE-HALL OF SOUTHEAST ASIA PTE. LTD., *Singapore*
WHITEHALL BOOKS LIMITED, *Wellington, New Zealand*

To Mary Lou

Contents

Preface

This book is designed as a text for a one-semester, nonmathematical optics course at the freshman or sophomore level of college. It has grown out of lecture notes for just such a course that I have taught at Florissant Valley Community College for the past eleven years. Although it was written with art majors in mind, it should be suitable for any nonscience majors: Students in my course who have successfully mastered this material include fine art, commercial art, graphics, fashion merchandising, theater, photography, and liberal arts majors.

Because of the intended audience a great deal of effort has gone into presenting reasoning without recourse to the mathematics that would ordinarily accompany such a physical science. There are only about a half dozen equations used in the body of the text, and those are all linear with at most four variables. Additional mathematical details are included in appendices.

The choice of topics covered has also been influenced by the intended audience. There is less material on optical instruments and physical optics than one would expect in a standard optics text, but more on lasers, holography, meteorological optics, and the psychology of vision. The geometrical optics chapter does have what some might consider an inordinate amount of material on the optics of conic sections, but the purpose is to prepare the reader for Tung H. Jeong's clever treatment of holography in terms of hyperbolic reflectors.

It is not necessary in a one-semester course to cover every chapter nor is it necessary to take them in order. The first four chapters form a basis for any of the others, which are largely independent of each other. For example, I have often taught the chapters in the order 1, 2, 3, 4, 8, 9, 10, 11, 6, 7, covering Chapter 5 at the end only if there is time. Certainly other teachers will have other preferences. Furthermore, the subheadings within chapters are designed to allow teachers to "fine tune" the text to their courses by omitting or adding specific topics.

I would like to take this opportunity to acknowledge the invaluable assistance of Jerry Thompkins of the Florissant Valley Physics Department, who helped with the photography, and the photographic and graphics units of Instructional Resources at the college, each of which lent a hand at critical times. Also I am indebted to the reviewers of the original manuscript for Prentice-Hall: Professor Donald D. Ballegeer, University of Wisconsin, Eau Claire, Wisconsin; Professor Joseph L. Aubel, University of South Florida, Tampa, Florida; and Professor Stanley H. Christensen, Kent State University, Kent, Ohio, whose suggested revisions were always good, if sometimes beyond the modest abilities of the author. In addition I must express my gratitude to two fine typists, Delores Orr and Jane Layton. Last, but not least, I would like to thank the many students who have wrestled with this material in my course; their struggle to understand has been the major inspiration for this work.

Gary Waldman

Part I

What Is Light?

Chapter 1

Early Ideas of Light

Since vision is the primary sense of human beings, we may be certain that people have wondered and speculated about light for many thousands of years, since long before there was any method of writing down those thoughts. Most early civilizations worshipped a sun god in some form, but in about 1370 B.C. Pharoah Akhenaton of Egypt introduced worship of a modified version of such a divinity that included the rays of sunlight. He saw the light from the sun as life-giving and had it clearly depicted that way in the Amarna style of Egyptian art.[1] In the Judeo-Christian tradition, the first chapter of the Bible depicts God's first act of creation as producing light. "And God said, Let there be light: and there was light."[2] The book of Genesis was probably not written down in its present form until 700 B.C. or later, but the tradition may go back much further.[3] By this date we are approaching the time of Greek civilization and with the Greeks, the first attempts at rational, nonreligious explanations of nature.

The Greek philosopher whose ideas about nature were most influential was Aristotle (384 B.C. to 322 B.C.) One reason that his ideas were accepted over such a long time was because he offered a complete and unified picture of the world. His theories of light, although they may seem strange to us today, were just a part of that overall picture. For Aristotle, the key to the nature of light was in transparent bodies, such a body being defined as anything "owing its visibility to the colour of something

[1]Cyril Aldred, *Akhenateu and Nefertiti* (New York: The Viking Press, 1973), pp. 12–20.

[2]The Holy Bible, *Authorized (King James) Version,* (Nashville: The National Publishing Co., 1972), p. 1.

[3]F. M. Cornford, "Pattern of Ionian Cosmogony," in *Theories of the Universe,* ed. Milton K. Munitz (New York: The Free Press, 1957), p. 29.

else; of this character are air, water and many solid bodies." He considered such things transparent because they contain a substance "also found in the eternal body which constitutes the uppermost shell of the physical Cosmos." Light, then is the activity of this almost divine substance. He did not think of light as a substance or even as moving but rather as the "presence of fire or something resembling fire in what is transparent."[4]

There the matter stood for over a thousand years. For although there may have been gains in the techniques of using light, there was no substantial advance in understanding the nature of light until that great flowering of inquiry in Europe known as the Renaissance shattered Aristotle's scheme of the universe. In the later sixteenth and early seventeenth century men such as Nicolaus Copernicus (1473–1543), Johannes Kepler (1571–1630), and Galileo Galilei (1564–1642) completely dismantled Aristotelian concepts in astronomy and mechanics and laid the foundations of modern science. One man of this period whose views are of particular interest to our study of theories of light was the French philosopher, René Descartes (1596–1650). Descartes, like Aristotle 2000 years earlier, tried to establish a unified world system that would explain all natural events. According to Descartes, all of space was filled with globules of a material which he called the "ether" that could transmit forces. A luminous body such as the sun caused a vortex or whirlpool in the ether. The outward centrifugal pressure from the vortex, transmitted through the globules pressing on one another, was light. In this theory, light had infinite speed: it was transmitted instantaneously. Of course, this hypothetical material called the ether was undetectable except insofar as it transmitted light and other forces such as gravity. Especially interesting was Descartes's view of colors as arising from rotation of the globules of the ether, with the most rapidly spinning particles giving rise to red sensations, the slowest giving blue, and particles with intermediate speeds giving orange, yellow, and green.[5] In the final analysis, Descartes's grand intellectual design proved far too ambitious an effort to be supported by the experimental evidence that was then available or obtainable. Furthermore he largely ignored the new emphasis on mathematical explanations in science due to Kepler and Galileo. Still, two of his ideas about light were to reappear in more successful theories: the first was the idea of light as a disturbance transmitted through the ether, and the second was the association of the different colors of the spectrum with different periodic motions of some kind.

QUESTIONS

1. What similarity can you detect between Akhenaton's and Aristotle's conceptions of light?

[4] Richard McKeon, ed., *Introduction to Aristotle* (New York: Random House, Inc., 1947), pp. 188–190.

[5] Sir Edmund Whittaker, *A History of the Theories of Aether and Electricity, Volume I: The Classical Theories* (New York: Harper & Row, Publishers, Inc., 1960), pp. 5–9.

2. What scientist is given primary credit for originating the idea of the ether?
3. In Descartes's theory what sort of motions are connected with color?
4. Compare the period of acceptance of Aristotle's "incorrect" theories with the period of acceptance of more modern theories of light dating from about 1800.

Chapter 2

The Classical Theories

2.1 THE CORPUSCULAR THEORY

The classical period of physics began in the late seventeenth century with the work of the English scientist Isaac Newton (1642–1727). Newton, one of the greatest geniuses of all time, established a complete science of motion and gravity that is still considered valid today except for the realms of very high speed or very small size: the calculations that send space probes to the moon or planets are based on Newtonian mechanics. In 1666 he also did some very important research into the nature of light that was not published until 1672. Still later (1705) he published a whole book entitled *Opticks.* Newton accepted Descartes's idea of an ether for the transmission of forces such as electricity or magnetism, but he could not accept the picture of light as a disturbance propagated through the ether. One reason he could not was because of the rectilinear propagation of light, in other words, because light travels in straight lines. He knew that disturbances in a medium, such as waves in water or sound in air, tend to bend around obstacles, whereas obstacles to light produce sharp shadows. This effect is illustrated in Fig. 2–1. In this respect, it is interesting to note that some small amount of illumination had already been discovered in the supposedly sharp shadows of opaque bodies. A Jesuit priest, Fr. Francesco Maria Grimaldi (1613–1663), had made the observation, which had been published posthumously in 1665.[1] Newton, however, considered this effect to be a kind of extended **refraction,** the bending of light when it passes from one medium to another.[2] This bending of

[1] Sir Edmund Whittaker, *A History of the Theories of Aether and Electricity, vol. 1: The Classical Theories* (New York: Harper & Row, Publishers, Inc., 1960), p. 13.

[2] Ibid., p. 20.

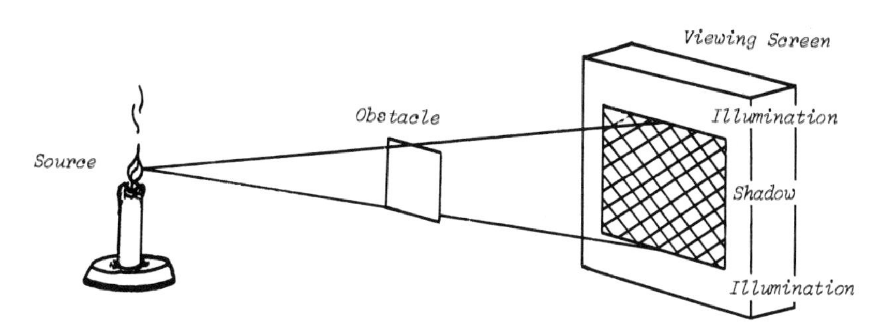

Figure 2-1 Newton's corpuscle evidence

light near an obstacle, named **diffraction** by Grimaldi, was certainly not nearly as pronounced as what could be observed for water waves and sound waves.

What then was Newton's concept of light? On the subject of light and in other scientific fields, he tried to refrain from making sweeping generalizations, but he did make it clear that he thought light to be the motion of some other substance through the ether, perhaps small particles or corpuscles, issuing from the luminous body.[3] Therefore, Newton's name came to be associated with the corpuscular theory of light. Streams of particles moving rapidly out from a luminous object in straight lines could easily explain the sharp shadows of opaque bodies.

2.2 THE WAVE THEORY

But still the idea of Descartes that light was a pressure or disturbance in the ether itself held its fascination for many scientists. The greatest champion of this view in Newton's own time was the Dutch scientist Christian Huygens (1629–1695). Huygens wrote a long paper in French summarizing his views in 1678, but it was not widely circulated until it was printed as a *Treatise on Light* in 1690. The evidence that convinced Huygens that light was a pressure wave moving through the ether was the fact that two beams of light whose paths cross are unaffected by the crossing but continue undeflected on their way.[4] The same result is easily seen with water waves, which can pass through each other and continue in their original direction, whereas any two streams of particles (imagine the jets from two water hoses) would deflect each other in crossing, as in Fig. 2-2. It has been suggested that it was quite natural for a Dutch physicist to rely upon waves as an explanation, since the Netherlands was a seafaring nation and one interlaced by canals, allowing natives to study wave phenomena daily from childhood.[5]

[3] Ibid., p. 19.

[4] Ibid., pp. 23–24.

[5] Ernst Mach, *The Principles of Physical Optics: An Historical and Philosophical Treatment,* trans. John S. Anderson and A. F. A. Young (New York: Dover Publications, Inc., 1926), p. 257.

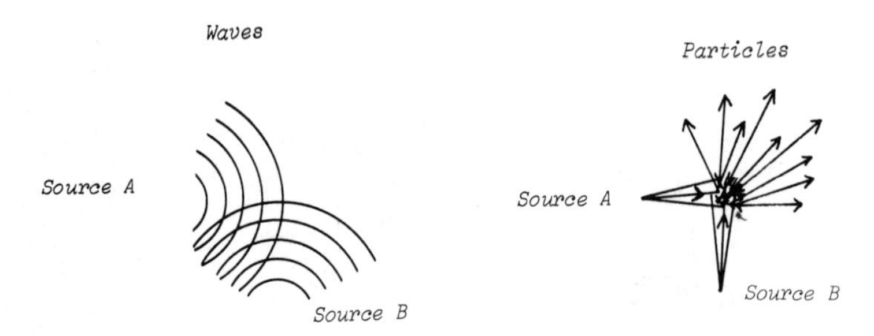

Figure 2–2 Huygens wave evidence

In his treatise, Huygens introduced a method of following the progress of a wavelike disturbance in a medium that was to prove most fruitful over the years. The method, now known as **Huygen's principle,** stems from his conviction that light is a succession of irregular pulses and is only detectable when a number of these elementary waves (or wavelets) unite in a larger wave. Huygens considered each point on a wave front or line of disturbance as a new source of wavelets and then added them to get the new wave front. Using this principle along with the mistaken idea of *irregular* pulses from the source, Huygens was able to show that his ether waves would give sharp geometrical shadows, indicative of rectilinear propagation.[6] His arguments now seem a curious mixture of the true and false, but the principle that each point on a wave front may be considered as a new source of waves was to prove instrumental much later in establishing the wave theory.

Thus we have from the very beginning of classical physics two opposing views of light: the wave theory and the corpuscular theory. For the whole eighteenth century (1700–1800) the corpuscular theory of light was accepted by most scientists, primarily because of the great authority of Newton, whose theories of mechanics were proving to be both profound and precise. This acceptance was not total, however, since a few scientists and philosophers still supported the wave theory. Among these wave advocates was Benjamin Franklin (1706–1790),[7] whose experiments in electricity and political revolution were his major claim to fame. With the great advantage of hindsight, we can see that eighteenth century science was deficient in experiments that could have provided reliable evidence for either theory of light.

In the next 50 years the situation became completely reversed, so that by 1850 the wave theory was as widely accepted as the corpuscular theory had been a century earlier. This reversal was primarily the work of two men, the Englishman Thomas Young (1773–1829) and the Frenchman Augustine Fresnel (1788–1827). Young, in particular, was a most remarkable man who made contributions to the fields of medicine, linguistics, mechanics, and optics. Trained as a medical doctor, between 1802 and 1804 he became the first person to conclusively demonstrate interference ef-

[6] *Ibid.,* p. 256.

[7] Whittaker, *A History of the Theories of Aether and Electricity,* p. 97.

fects in light experiments, effects that could only be explained by the wave theory. To understand interference effects we need to study wave motion more carefully.

2.2.1 Wave Description

A side view of a wave, such as a water wave or a wave on a string, is shown in Fig. 2–3. Here, the whole wave must be imagined as moving left to right. In this type of wave, called a **transverse wave,** the particles of the medium move up and down, that is, perpendicular to the motion of the wave. Huygens did not think of light as a transverse wave but rather as a longitudinal wave, in which the particles of the medium move back and forth along the direction of the wave motion.[8] By about 1820 Young and Fresnel were convinced that light waves must be transverse because of **polarization** phenomena, which we shall consider later.[9] The basic quantity used to describe the wave is the **wavelength,** symbolized by the Greek letter λ (lambda). Wavelength is defined as the distance between successive crests; it is also the distance between successive troughs. As a matter of fact, the wavelength can be taken from any point to the next corresponding point at which the wave is undergoing the same motion. Thus a uniform, regular wave is made up of repetitions of identical segments of one wavelength. Wavelength is measured in length units such as meters. The wave is also described by the **amplitude,** which is the height of the crest above (or depth of the trough below) the zero line. Amplitude is indicated by A, and in a water wave or a wave on a string would be measured in length units. Since the wave is considered to be moving, the crests and troughs all have a wave speed v, measured in velocity units such as meters per second (m/s).

One other important quantity describing the wave is more difficult to visualize than those just discussed. That quantity is called **frequency** f, defined as the number of waves (or crests or troughs) passing a fixed point in the path of the wave per unit time. It has units of cycles per second (cps) or in more modern terms, hertz (Hz).

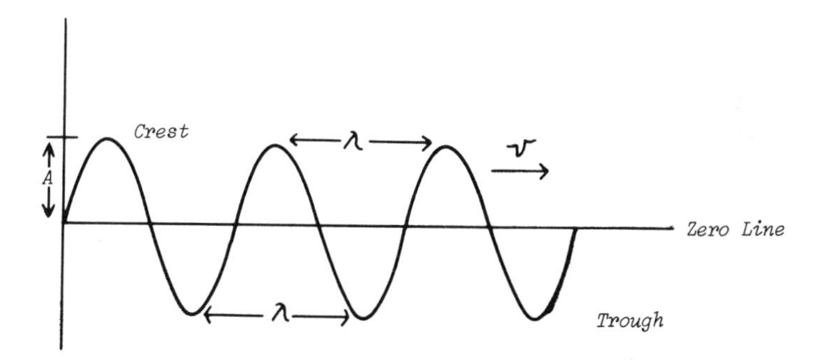

Figure 2–3 Waves

[8] Mach, *Principles of Physical Optics,* p. 188.

[9] *Ibid.,* pp. 201–203.

Thus, if you could sit and count waves as they pass, and you counted seven going by in 1s, the frequency of the wave would be $f = 7$ cps $= 7$ Hz.

Wavelength λ, speed v, and frequency f are of necessity related to each other just because of the way they are defined. For any type of wave, we must have speed = wavelength × frequency.

$$v = \lambda f \tag{1}$$

This equation expresses the fact that if f waves pass in 1 s and each is λ meters long, the total wave travel in 1 s must be λf.

Example:

If International A of the musical scale is a wave of frequency 440 Hz and the speed of sound in air is 340 m/s, what is the wavelength of this sound wave?

$$\text{Since } v = \lambda f$$

$$\lambda = \frac{v}{f} = \frac{340}{440} = 0.773 \text{ m}$$

2.2.2 Superposition of Waves

Interference can take place whenever two waves are present in the same medium at the same time. Then a principle of superposition applies, which states that the displacement of the medium at any point is equal to the sum of the displacements of each of the two waves taken separately. However, it is crucial to remember that the displacement of the medium in Fig. 2–3 can be either up or down, and the sum of an up and a down displacement means subtracting one from the other. Superposition of two waves is shown in Fig. 2–4, with the two waves shown separately above and the resultant wave shown below. You can see that at some points the two displacements give a greater result than either alone or at other points the two tend to cancel each other.

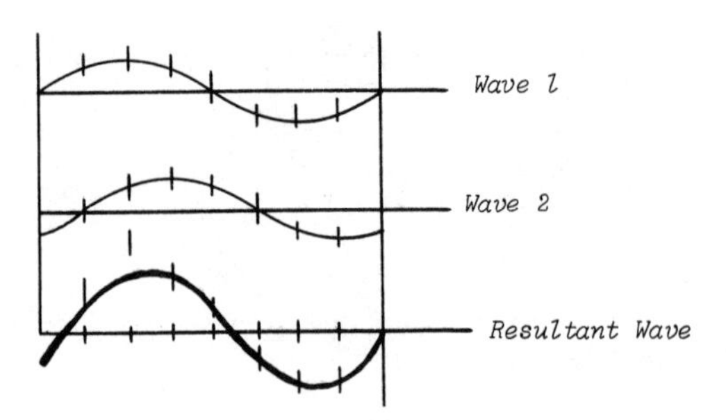

Figure 2–4 Superposition of waves

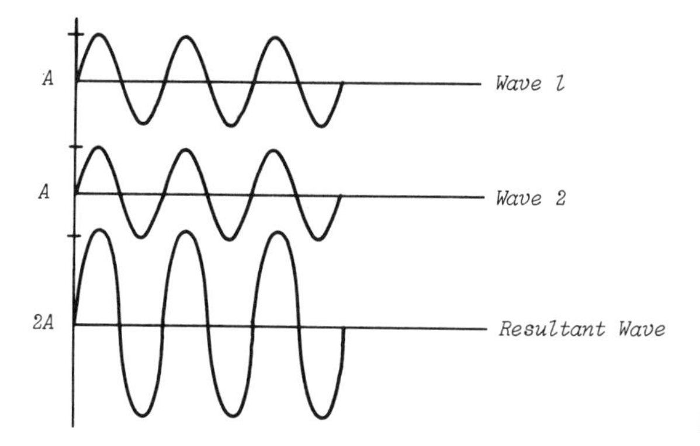

Figure 2–5 Constructive interference

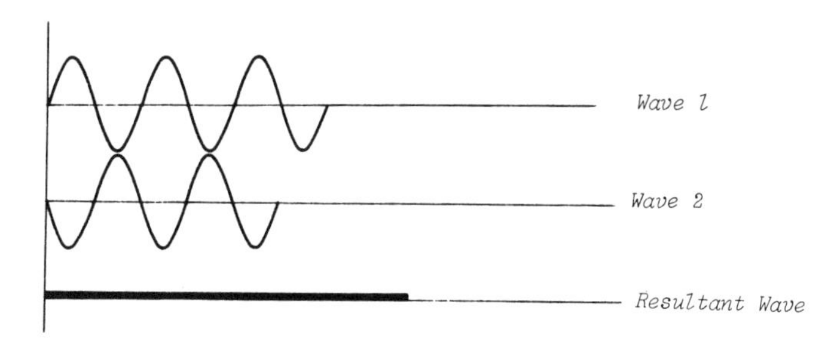

Figure 2–6 Destructive interference

Two situations of great importance can arise when the two interfering waves have the same amplitude and wavelength. The first situation, which occurs when the two waves are in phase, or aligned crest to crest and trough to trough, is called **constructive interference.** In this case the resultant wave has the same wavelength but twice the amplitude, as shown in Fig. 2–5. The opposite situation arises when the two interfering waves are aligned crest to trough (said to be 180° out of phase). Here the upward displacement due to one wave everywhere just cancels the downward displacement due to the other; the two waves together produce no wave. This rather surprising result, called **destructive interference,** is shown in Fig. 2–6.

2.2.3 Young's Double-Slit Experiment

Thomas Young's most convincing interference experiment demonstrated this effect of two light sources interfering destructively to produce darkness, a result wholly unexpected on the basis of the corpuscular theory. Young's double-slit experiment used a point source of monochromatic (single wavelength) light and split the light into two parts by passing it through two slits in an opaque screen. The light from the two slits was then viewed on a screen opposite the point source. Of course, the particle theory of light would predict just two spots of light on the viewing screen, at positions on straight lines from the source through the two slits, as shown in Fig. 2–7. In

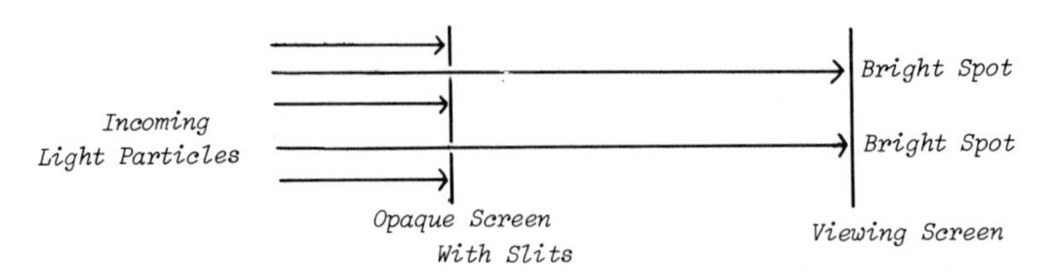

Figure 2–7 Double-slit: particles

the wave theory, however, Huygens' principle asserts that each slit would act as a new source of waves, which would then overlap and produce an interference pattern on the screen. In some places crests from the two slits would always arrive together, giving constructive interference and a bright spot. At intermediate positions, crests from one slit would always arrive at the same time as troughs from the other, giving destructive interference and darkness. Thus an interference pattern of many bright and dark spots is predicted, as shown in Fig. 2–8.

Young's performance of the double-slit experiment not only confirmed the wave theory of light but also allowed him to measure its wavelength. We now recognize that the different colors of the spectrum correspond to different wavelengths, with the shortest visible wavelengths producing the perception of violet and the longest the perception red. All visible wavelengths are quite small on a human scale, being measured in nanometers (billionths of a meter abbreviated nm). We have

380 to 450 nm → violet
450 to 490 nm → blue
490 to 560 nm → green
560 to 590 nm → yellow
590 to 630 nm → orange
630 to 760 nm → red

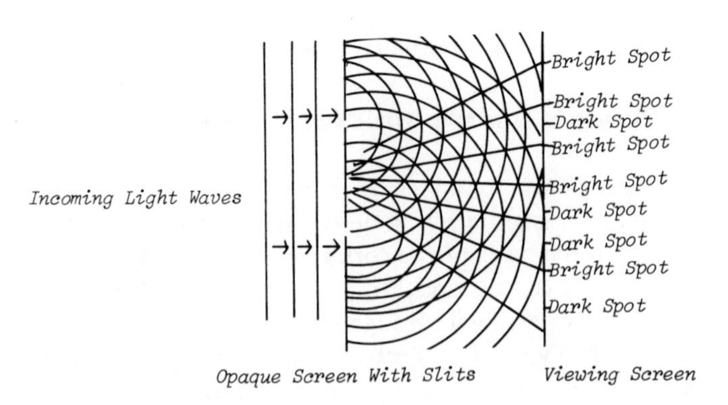

Figure 2–8 Double-slit: waves

Of course, the color perceptions merge gradually into one another with changing wavelength, and both ends fade into darkness. Often the visual spectrum is nominally taken to be 400 nm to 700 nm. These wavelengths are all less than a millionth of a meter, and this short length may be taken as a major reason that the wave nature of light was not discovered earlier.

2.2.4 Speed of Light Measurements

Other important measurements were being made in the nineteenth century on the speed of light. Descartes had thought that light was transmitted instantaneously, but many scientists disagreed. Galileo was one of the earliest to design and attempt an experiment. His experiment involved two observers, each with a shuttered lantern, stationed miles apart at night. One observer would uncover his lantern and start timing. When the second observer saw the flash from the first lantern, he was to uncover his. The first observer stopped timing when he saw the light from the second lantern. As you might imagine, the speed of light is too great to be measured by this crude method.

In 1675, the Danish astronomer Olaf Roemer (1644–1710) made an astronomical measurement of the speed of light, using apparent variations in the motions of Jupiter's satellites. This was the first time a value for the speed was obtained. In 1727 another astronomical determination of the speed of light was obtained by the English astronomer James Bradley (1692–1762), who used an effect called **stellar aberration.** Bradley's result at least roughly confirmed the earlier measurements using Jupiter's moons.

The first terrestrial measurements of the speed of light were not made until the nineteenth century and were instrumental in confirming the wave theory. The French scientist Armand Fizeau (1819–1896) in 1849 measured the speed of light with the ingenious apparatus shown in Fig. 2-9. Here S is a source that produces a beam of light, which is reflected off the half-silvered plate P, past a toothed wheel W, to a mirror M, 8 km away. The beam is directed so that it will pass through a gap in the wheel but be blocked by a tooth. The light is returned by the mirror along exactly the same path and passes through the half-silvered plate to the observer O. When the toothed wheel is spun by a motor at relatively low speed, the light can still be seen by the observer because it can travel all the way to M and back to W before the next tooth interrupts the beam. But if the speed of the wheel is continually increased, a point will

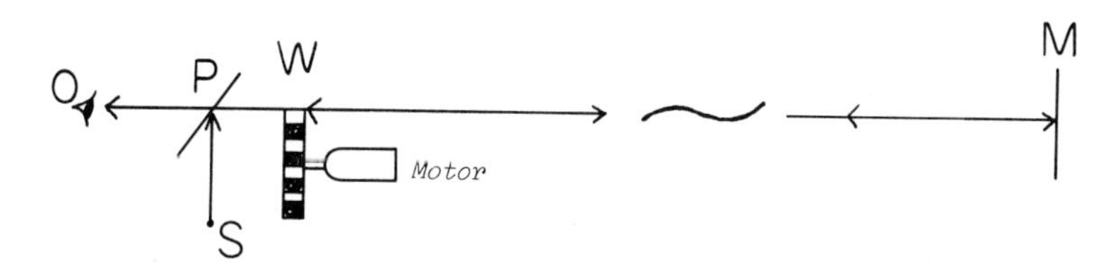

Figure 2-9 Fizeau apparatus

be reached at which the light seen by the observer goes out. At this point the light that is admitted by the leading edge of each gap is just intercepted by the leading edge of the next tooth. Therefore knowledge of the wheel speed when this extinction takes place allows one to calculate the time for light to travel 16 km and from that to calculate the speed of light. Similar experiments over the past century have established the value $c = 3 \times 10^8$ m/s for the speed of light in a vacuum (nearly the same value as in air). For rough calculations in English units, we sometimes use 10^9 ft/s or 1 ft/ns (ns stands for nanosecond, one-billionth of a second).

A crucial measurement for the wave theory of light was that of the speed of light in a relatively dense medium, such as water or glass. In order to explain the observed facts of **refraction,** that is, the bending of light when it passes from one medium to another, the corpuscular theory had to assume that light speeded up when it went from air into water while the wave theory had to assume just the opposite. In 1850 the French scientist Jean Bernard Foucault (1819–1868) succeeded in measuring the speed of light in water and found it to be less than the value in air. We may consider this measurement as the final nail in the coffin of the classical corpuscular theory of light.

A knowledge of both the speed and wavelength of light allows us to calculate the frequency of light. Equation 1 still applies, but $c = 3 \times 10^8$ m/s must be used for the speed.

Example:

What is the frequency of green light with $\lambda = 500$ nm?

Here $c = \lambda f$

and $f = \dfrac{c}{\lambda}$

with $\lambda = 500\,\text{nm} = 5 \times 10^{-7}\,\text{m}$

$$f = \frac{3 \times 10^8}{5 \times 10^{-7}} = 0.6 \times 10^{15}\ \text{Hz}$$

$$f = 6 \times 10^{14}\ \text{Hz}$$

This example shows what tremendously large frequencies are typical of light. Six hundred trillion waves per second pass any point in the path of this light wave!

2.3 ELECTROMAGNETIC WAVES

Apart from establishing the wave nature of light, classical physics contributed one other great breakthrough to the understanding of light. This breakthrough was contained in the work of the English physicist James Clerk Maxwell (1831–1879). Max-

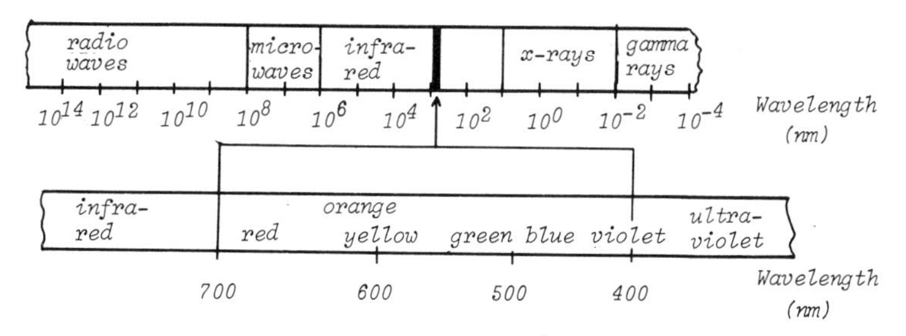

Figure 2–10 Electromagnetic spectrum

well was one of a long line of English scientists who worked to explain electric and magnetic effects in terms of the mechanics of that mysterious substance, the so-called ether. In theoretical work on the nature of electric and magnetic forces, Maxwell reached the conclusion in 1861 that an electric charge that was changing its velocity (in other words, accelerating) should produce a disturbance in the ether that travels outward from the source with a speed of 3×10^8 m/s.[10] Since the speed Maxwell found was calculated solely from constants of electricity and magnetism, the fact that it was equal to the already measured speed of light suggested strongly that light was just such a disturbance. To produce a regular, uniform wave the electric charge would have to vibrate back and forth; for light frequencies such a charge would have to oscillate several hundred trillion times a second.

We now know that Maxwell was correct: light is part of a much larger class of waves known as electromagnetic waves. There is no limit to how short the wavelengths of electromagnetic radiation can be, nor to how long. Light is distinguished simply by the fact that the human eye responds only to those electromagnetic waves with wavelengths between 400 nm and 700 nm. Electromagnetic radiation with wavelengths just greater than 700 nm is known as infrared radiation. At the opposite end of the visible spectrum is ultraviolet radiation, with wavelengths just below 400 nm. Figure 2–10 shows the whole electromagnetic spectrum.

QUESTIONS

1. Name an early (pre-1700) classical physicist advocating a corpuscular theory of light and describe his supporting evidence.
2. Name an early (pre-1700) classical physicist supporting a wave theory of light and describe his supporting evidence.
3. What is Huygens' principle?
4. What scientists were foremost in establishing the wave nature of light in the early nineteenth century?

[10] Whittaker, *A History of the Theories of Aether and Electricity,* pp. 250–253.

5. What is the approximate wavelength of blue light? Of orange light?
6. Who made the first astronomical measurement of the speed of light? The first terrestrial measurement?
7. If you wish to turn out the light in your room and jump into bed before it gets dark, how much time do you have if the lamp is 7 ft from the bed? Assume the lamp stops emitting light as soon as it is switched off.
8. What scientist first identified light as a kind of electromagnetic wave?
9. What is the frequency of violet light with a wavelength of 400 nm?
10. What would electromagnetic radiation with a wavelength of 950 nm be called?

Chapter 3

Modern Theories

By the beginning of the present century most scientists were satisfied that the basic nature of light was understood. In fact, many scientists thought that the whole physical world was essentially explained by the theories of the time. The sciences of heat and thermodynamics had been reduced to the mechanics of many small particles (atoms and molecules), viewed statistically. Electricity, magnetism, and optics had all been theoretically combined, and their reduction to the mechanics of the ether was confidently expected by many. If physics was the queen of the sciences, mechanics, based on Newton's laws, was the queen of physics. There were just some calculation details to work out and a few theoretical discrepancies to explain. But those few discrepancies proved to be the wedge that split classical physics wide open and paved the way for what we call modern theories. The program of reducing all science to Newtonian mechanics proved after all to be a dead-end road.

One of the discrepancies that physicists faced around 1900 involved the emission of light, as well as other electromagnetic radiation, by solids. They wished to be able to explain the details of the spectrum emitted by applying the classical theories of mechanics and electricity to the myriads of atoms that make up the solid. Actually, scientists wished also to explain the emission spectrum of gases, which were quite different from those of solids. Both solids and gases emit light when heated to high temperature; however, when that light is viewed through a spectroscope, an instrument that spreads light out into its constituent wavelengths, the solid spectrum is a continuous band of color from red to violet while the gas spectrum consists only of several different colored lines. Therefore we speak of **continuous spectra** and **line spectra.** These results imply that solids emit all wavelengths while gases emit only certain discrete wavelengths. Furthermore, different gases emit different sets of wavelengths.

3.1 BLACKBODY RADIATION

The continuous spectrum emitted by a hot solid is often called **blackbody radiation** because a perfectly black object (one that absorbs all the radiation falling on it) is easiest to treat theoretically and most solids emit radiation in a manner similar to the way blackbodies do. In this type of spectrum, there is some power at all wavelengths, but there is not equal power at all wavelengths. We can characterize a spectrum such as this one by drawing a graph of power emitted at each wavelength versus the wavelength. Generally we get a shape like that shown in Fig. 3–1.

At any given temperature, say T_1, the power radiated at very short wavelengths is quite low, as shown by the height of the curve in the figure being nearly zero. As one looks at longer and longer wavelengths, the curve rises until it reaches a maximum at the wavelength designated λ_{m1}, which is called the **wavelength of maximum emission** because, as the graph indicates, more power is emitted at that wavelength than at any other. For wavelengths longer than that of maximum emission, the radiated power decreases in the long tail seen on the graph, eventually approaching zero again for very long wavelengths.

When the temperature of a blackbody is raised, say from T_1 to T_2, two changes may be noticed in the spectrum. First, since the whole curve is higher, more power is emitted at all wavelengths. The second change is seen in the shift of the peak of the curve to a shorter wavelength. In other words, the wavelength of maximum emission changes from λ_{m1} to λ_{m2}. You can easily observe this shift in wavelength by watching the heating element of a toaster or electric range just after it is turned on. The first visible radiation is a dark red, and then as the temperature increases, the color changes to an orange-red. This color change is a result of the steeply sloped left-hand edge of the radiation curve shifting through the red end of the visible spectrum. If the

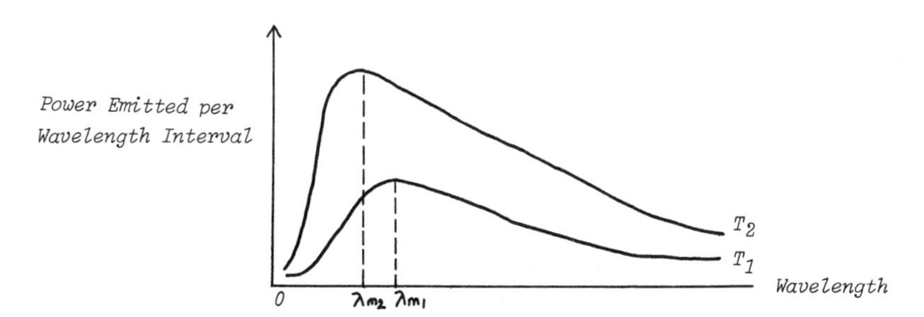

Figure 3–1 Blackbody spectrum

element became as hot as the tungsten filament of an incandescent lamp (which it does not), its color would appear yellowish white.

Actually the element was radiating power in the form of electromagnetic waves even before it was turned on. Every object above absolute zero (0 K) in temperature shows such radiation. But for the temperatures usually found on the earth's surface the radiation curve is so far out in the infrared that there is no perceptible radiation in the visible region. For example, at the temperature of the human body, $\lambda_m \cong 9400$ nm; you are presently radiating power, but at wavelengths around 9000 nm and longer. For infrared wavelengths such as these, it is convenient to use as the length unit the micrometer (μm), one millionth of a meter. In these terms

$$1\,\mu\text{m} = 1000\,\text{nm} = 10^{-6}\,\text{m} \tag{1}$$

Then the visible region is 0.4 μm to 0.7 μm, and for a temperature of 100° F ≅ 38°C, $\lambda_m = 9.4\,\mu$m. For a solid at the boiling point of water (212°F = 100°C), $\lambda_m = 7.8\,\mu$m, and to shift λ_m down to the long end of the visible at 0.7 μm, the temperature must be raised to 7000°F ≅ 3870°C.

Every attempt to explain the shape of the blackbody radiation curve of Fig. 3–1 through the use of classical physics ran into serious problems. Most notably, in the years 1900 to 1905 the English scientists Lord Rayleigh (1842–1919) and Sir James Jeans (1877–1946) used classical ideas of electromagnetism and statistical mechanics to derive a theoretical curve for blackbody radiation. This curve, sometimes called the Rayleigh-Jeans law, is shown, along with the corresponding experimental blackbody curve, in Fig. 3–2.

As can be seen from the figure, the Rayleigh-Jeans law fits the facts fairly closely at long wavelengths (low frequencies), but diverges widely from experimental results at short wavelengths (high frequencies). In fact, the Rayleigh-Jeans curve approaches infinity as the wavelength approaches zero. This divergence of the Rayleigh-Jeans law was called the "ultraviolet catastrophe" because it occurs at short wavelengths and there seemed to be no way for classical physics to avoid such a nonsensical result.

3.1.1 The First Quantum Theory

In 1900 another scientist in another country was proceeding along different lines to arrive at results that not only correctly predicted blackbody radiation but also

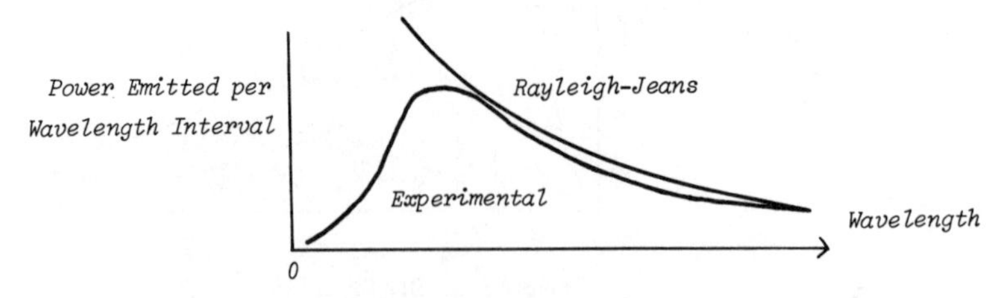

Figure 3–2 Blackbody spectrum and Rayleigh-Jeans law

represented the beginning of a new, nonclassical physics known as **quantum mechanics.** The German physicist Max Planck (1858–1947) was examining the blackbody radiation curve to see what assumptions would be necessary to correctly derive it. He found that if he abandoned some classical ideas about the emission of electromagnetic energy by the atoms of a solid, but instead assumed that they radiated energy in tiny chunks, called **quanta** (singular, quantum), he could derive a formula that fit the experimental curve very accurately at all wavelengths. Each quantum emitted had to have an energy proportional to the frequency of the electromagnetic radiation. If E stands for the energy of one quantum, then

$$E = hf \tag{2}$$

where h, which has the value 6.62×10^{-34} Js is called **Planck's constant.** The J in the units of h stands for the energy unit joule, so that when units of h are multiplied by units of f and s^{-1} (reciprocal seconds) energy units are obtained. The joule is a reasonably sized energy unit on the human scale, being approximately equal to the energy a person would use in lifting 1 kg (2.2 lb) 10 cm (4 in.) Therefore the energy of an individual quantum hf, even if f is a high frequency, will be incredibly small on the human scale of things. That is one reason such "graininess" in the energy of electromagnetic waves was not discovered until 1900. Furthermore, the graininess is more pronounced for high frequencies than for low frequencies, which is why the Rayleigh-Jeans law fits at one end of the spectrum but falls victim to the ultraviolet catastrophe at the other.

Example:

What is the energy of one quantum of the green light of the last example?

We had found for $\lambda = 500$ nm

$f = 6 \times 10^{14}$ Hz

Therefore $E = hf = (6.62 \times 10^{-34})(6 \times 10^{14})$

$E = 39.6 \times 10^{-20}$ J $= 3.96 \times 10^{-19}$ J

At the time even Planck himself was not happy about the implications of his quantum hypothesis. Classical physics had always been successful up to that time and no one wanted to abandon it for uncertain new ideas. Planck and others thought that a way still might be found to explain blackbody radiation classically and that the quantum hypothesis would then be seen to have been just a temporary, ad hoc explanation that worked out by luck. But the quantum theory did not fade away in that manner. Instead, it proved to be more and more useful for explaining phenomena that had baffled classical physics. One such phenomenon was the photoelectric effect.

3.2 PHOTOELECTRIC EFFECT

An explanation of the **photoelectric effect** had been eluding scientists since it was first discovered in 1887 by Heinrich Hertz (1857–1894), the same scientist who had first experimentally confirmed Maxwell's prediction of electromagnetic waves and after whom the frequency unit is named. The effect is basically the release of negative electric charge by some metals when their surfaces are irradiated with electromagnetic waves of high frequency; some metals show the effect when irradiated with visible light, but generally ultraviolet waves are required. Only later was it learned that the charge was released in the form of elementary negative charges called **electrons** (electrons themselves were not discovered until 1897). One aspect of the photoelectric effect that was puzzling was the fact that increasing the energy of the incoming light beam by making it more intense did not increase the energy of the individual released electrons; instead it produced more electrons. Also puzzling was the fact that increasing the frequency of the incoming radiation did increase the energy of the individual released electrons. Perhaps most puzzling was the classical calculation that it would take minutes to days for an electron in the metal to store up enough energy from the incoming waves to be released from the surface at all,[1] whereas experiments showed that photoelectric emission was practically instantaneous.

In 1905 Albert Einstein (1879–1955) gave an explanation of the photoelectric effect, based on the idea of quanta of electromagnetic radiation, which was simple, accurate, and convincing. His basic assumption was that electrons in the metal do not absorb energy from the electromagnetic waves in a continuous manner, but rather they absorb chunks or quanta of exactly the same size as Planck had theorized, $E = hf$. The energy is already concentrated into quanta when absorbed and so the electrons do not have to wait to store up energy. Each incoming quantum of electromagnetic energy releases one electron. Increasing the intensity of the incoming radiation increases the number of quanta but not their individual energies, and so the number of emitted electrons is increased. Increasing the frequency of the radiation increases the energy of each quantum, according to Eq. (2), and thus increases the energy of each electron emitted.

After 1905 it was no longer possible to ignore the quantum theory or to reasonably hope that it would go away. This led to a real dilemma concerning the nature of light. If light is emitted in quanta in blackbody radiation and absorbed in quanta in the photoelectric effect, does it really consist of particles instead of waves? In modern theories such light particles are known as **photons.** We shall see that additional successes of the quantum theory eventually forced scientists to face squarely this question of the dual (wave and particle) nature of light. But first, we need to examine some of those additional successes.

[1] F. K. Richtmyer, E. H. Kennard, T. Lauritsen, *Introduction to Modern Physics,* 5th ed. (New York: McGraw-Hill Book Company, Inc., 1955), p. 98.

3.3 THE NUCLEAR ATOM

Advances in our knowledge of the nature of atoms soon produced another dilemma for physicists. In 1911 the English physicist Ernest Rutherford (1871–1937) offered a new model of the atom based on the results obtained from the scattering alpha particles (subatomic particles produced by radioactive substances) by gold atoms in a thin foil. The scattering experiments indicated that the positive electric charge and almost all the mass of an atom is concentrated in a very small central nucleus; the electrons of the atom revolve around the nucleus in orbits, much as the planets in the solar system orbit the sun. This model is shown in Fig. 3–3.

Scientists had hoped that a correct model of the atom would provide an explanation of the line spectra of gases, which we discussed briefly earlier. Since the atoms of a gas do not interact with one another except during the very brief intervals when they are colliding, it was believed that the line spectrum of a gas was determined by the internal structure of the individual atoms rather than by collective or statistical properties of the atoms as a group, as in the case of blackbody radiation. This belief was reinforced by the fact that different elemental gases showed different line spectra and therefore should have atoms with different internal structure, as required by the known results of chemistry. In addition, a cool gas will absorb exactly the same wavelengths from white light that it emits when heated.

Rutherford's model of the atom, however, was of little help in understanding the line spectra of gases. According to the classical theory of electromagnetic waves, any electron traveling in a circular orbit should radiate continuously at the same frequency with which it revolves around the nucleus. Such an electron, because it is radiating away energy, should lose energy and spiral into the nucleus, increasing its frequency of revolution as it does so. Therefore, not only was the Rutherford atom theoretically unstable, but it also should emit radiation all during its existence and it should emit all frequencies (or wavelengths) in a very short time interval. Yet the model had apparently been confirmed by scattering experiments.

3.3.1 Bohr Model

Again it was quantum theory that came to the rescue. In 1913 the young Danish physicist Niels Bohr (1885–1962) offered an explanation of the Rutherford atom that

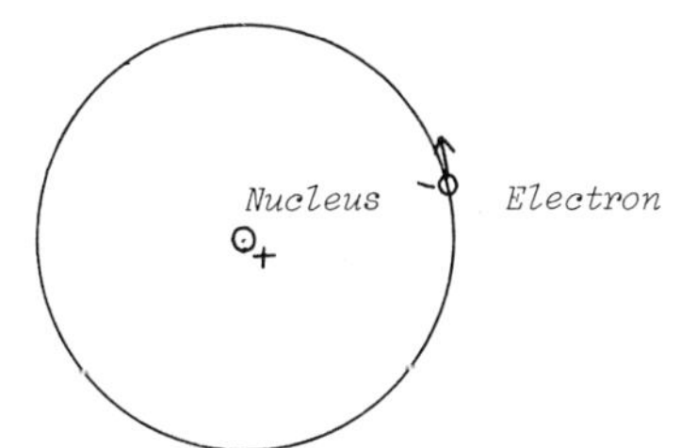

Figure 3–3 Rutherford model of the atom

correctly predicted the spectral lines of hydrogen by incorporating the ideas of quantum theory. Hydrogen is the lightest element and has a relatively simple line spectrum, so it has always been thought to have the simplest atoms and has always been the first subject of theoretical analysis. Bohr's theory was basically classical but incorporated with quantum ideas at certain key points. His assumptions contrary to classical physics were:

1. The electron can revolve only in certain discrete orbits around the nucleus, not in any orbit at any distance.
2. The electron does not emit or absorb electromagnetic radiation while revolving in these circular orbits, but only when it jumps between orbits.
3. The electron emits or absorbs one quantum of electromagnetic radiation when it jumps between orbits.

Thus in the Bohr model of the atom each different discrete orbit has a different radius and corresponds to a different energy for the electron, the larger orbits corresponding to higher energies. Because whatever energy the electron loses when it jumps down in orbit (or gains when it jumps up) must be exactly the energy of one photon emitted (or absorbed), we can get a formula for the frequencies that the hydrogen atom may emit or absorb. Suppose the electron falls from orbit 2 to orbit 1; then the energy lost by the electron is $E_2 - E_1$, where E_2 is its energy in orbit 2 and E_1 is its energy in orbit 1. The energy of the photon is given by Eq. (2), so that

$$hf = E_2 - E_1 \tag{3}$$

There are possible jumps from orbits 2, 3, 4 . . . to 1, from orbits 3, 4, 5 . . . to 2, and so on. Each jump downward gives one emission line in the spectrum and each jump up gives an absorption line with the same frequency. The situation is shown in the diagram of Fig. 3–4.

It should be clear from Eq (3) and the subsequent discussion that if one knew the orbital energies (E_1, E_2, E_3 . . .) for the hydrogen atom, it should be possible to calculate the possible frequencies of the spectral lines. That is exactly what Bohr did, and the close agreement with experimental results that he obtained led most other scientists to readily accept his theory.

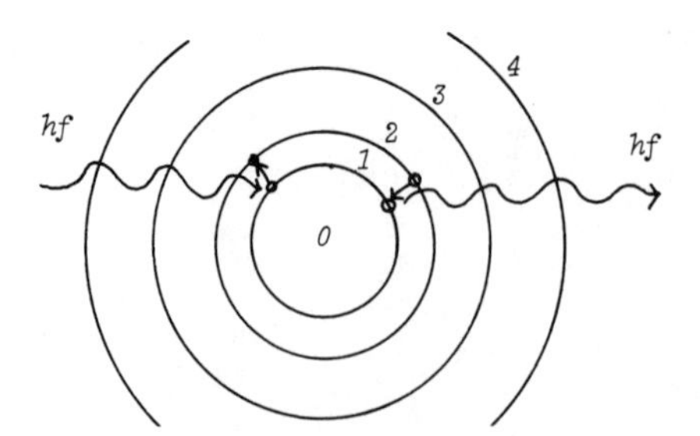

Figure 3–4 Emission and absorption in Bohr model

Example:

If the energy of the fourth Bohr orbit is $E_4 = -1.36 \times 10^{-19}$ J and that of the second Bohr orbit is -5.43×10^{-19} J, (the negative signs indicate that the electron is bound to the atom and not free) what frequency of radiation is emitted when the electron jumps from the fourth to the second orbit?

We have $hf = E_4 - E_2$

Then $f = \dfrac{E_4 - E_2}{h}$

$$f = \frac{-1.36 \times 10^{-19} - (-5.43 \times 10^{-19})}{6.63 \times 10^{-34}}$$

$$= \frac{(5.43 - 1.36) \times 10^{-19}}{6.63 \times 10^{-34}} = \frac{4.07 \times 10^{-19}}{6.63 \times 10^{-34}}$$

$$f = 0.614 \times 10^{15} = 6.14 \times 10^{14}\ \text{Hz}$$

The wavelength can also be found because we know from the previous section

$$c = \lambda f$$

$$\lambda = \frac{c}{f} = \frac{3 \times 10^8}{6.14 \times 10^{14}} = 0.489 \times 10^{-6}\ \text{m}$$

$$\lambda = 489\ \text{nm}$$

blue-green light!

It should be briefly noted here how Bohr established the radii and energies of his discrete orbits. He realized that the units of Planck's constant h were the same as the units of angular momentum, a quantity often used in classical physics to describe the motion of an object orbiting as do the electrons in the Rutherford atom. For a point mass m traveling at a constant speed v in a circle of radius r the angular momentum is mvr (mv, the mass times the velocity of an object, is called its linear momentum.) Bohr theorized that the electron in an atom might just have whole numbers of a basic chunk of angular momentum which was $h/2\pi$ ($\hbar$ is the symbol often used for $h/2\pi$). In other words, the electron could have an angular momentum of $\hbar$ or $2\hbar$ or $3\hbar$ and so on, but no value in between: $\hbar$ would represent a basic quantum of angular momentum. Therefore this hypothesis is called **quantization of angular momentum** and gives an equation which leads to only certain allowed orbital radii and certain allowed energies:

$$mvr = n\hbar,\ n = 1, 2, 3 \ldots \tag{4}$$

In this equation, if $n = 1$, the electron is in the first Bohr orbit, has an angular momentum of $\hbar$, and has an energy E_1. If $n = 2$ the electron is in the second Bohr orbit, has an angular momentum $2\hbar$ and an energy E_2, and so on up to higher values of n. Although we will not use this equation for any calculations, we will have occasion to refer back to it in discussing later concepts.

Bohr's idea of the quantization of angular momentum and his assumptions about the electron, discussed earlier, were ad hoc assumptions, not firmly connected to the rest of physics. Even by 1920 there was no certain path through quantum theory that scientists could follow to work out any problem. Most advances, like Bohr's, were combinations of classical physics with a quantum hypothesis thrown in somewhere as an inspired guess. Yet certainly the quanta and the photons were there, as real in their way as electromagnetic waves.

Additional evidence came in 1922 from the American physicist Arthur Holly Compton (1892–1962). In explaining the scattering of x-rays from electrons, he showed that the problem could be treated as a collision of particles: photons and electrons. The wave-particle duality of light (and all electromagnetic radiation) was obviously not about to go away. On the one hand, there were the interference experiments, such as Young's double-slit experiment, which showed that light has wave properties, while on the other hand there were the photon experiments, such as the photoelectric effect and the Compton effect, which showed that light has particle properties.

3.4 MATTER WAVES

Clearly what was needed was a unified theory that could be applied to any quantum problem, and one that would shed light on the wave-particle dilemma. Such a unified theory was developed between 1925 and 1930, and it was based on a brilliant insight by the French scientist Louis de Broglie (1892–), published in 1924. He suggested that since light, which classical physics had considered to be waves, sometimes behaved like particles, perhaps electrons and other particles sometimes behaved like waves. Reasoning by analogy with light, he derived a wavelength for the wave associated with a particle of mass m traveling at speed v. This de Broglie wavelength is

$$\lambda = h/mv \qquad (5)$$

In other words, Planck's constant divided by the momentum of a particle gives the wavelength of the associated wave. Because Planck's constant is so small, large objects such as a thrown rock have an immeasurably small wavelength—their wave properties cannot be detected. But electrons at typically encountered speeds have a small enough momentum to give a detectable wavelength.

Example:

Find the de Broglie wavelength of an electron ($m = 9.1 \times 10^{-31}$kg) traveling at 2×10^6 m/s (even in low-voltage, low-energy experiments, electrons typically travel at 1 million m/s or more):

$$\lambda = \frac{h}{mv}$$

$$\lambda = \frac{6.63 \times 10^{-34}}{(9.1 \times 10^{-31}) \times (2 \times 10^6)} = 3.64 \times 10^{-10}\text{ m}$$

$\lambda = 0.364$ nm about the same as x-rays

Small, but large enough to detect.

Example:

Find the de Broglie wavelength of a 20g = 0.02 kg bullet traveling at 500 m/s:

$$\lambda = \frac{h}{mv}$$

$$\lambda = \frac{6.63 \times 10^{-34}}{0.02 \times 500} = \frac{6.63 \times 10^{-34}}{10} = 6.63 \times 10^{-35}\text{ m}$$

Far too small to be detected!

By 1930 a unified quantum theory, called **wave mechanics** and based on de Broglie's "matter waves," had been completed. Subsequent successes of the theory more than justified de Broglie's bold new concept. For our purposes, we need not examine the later successes, but it is of interest to consider how de Broglie's ideas clarified the models of atoms and light that were current in 1924. Electron waves very naturally led to a condition for stable orbits in the hydrogen atom. If waves are traveling in a circular path around the nucleus, then the only stable orbits should be ones such that the waves exactly close on themselves in going once around. In that case, each time the wave goes once around it reinforces itself through constructive interference. On the other hand, if the wave does not exactly close on itself it will give some partial destructive interference in each trip around and cancel itself out in many trips. Figure 3–5 illustrates the stable configurations.

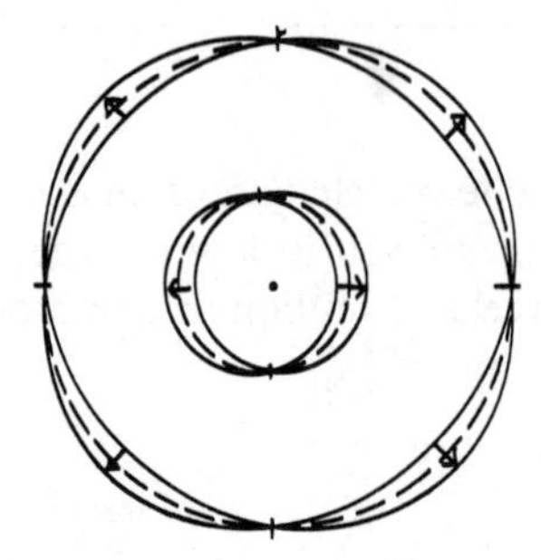

Figure 3–5 Electron orbits and waves

Referring to Fig. 3–5 and our definition of wavelength in Chapter 2, we can set down a mathematical condition for a stable orbit: The circumference of the circular orbit must be an integral number of wavelengths.

$$2\pi r = n\lambda, n = 1, 2, 3 \ldots \tag{6}$$

Next we substitute for λ, the wavelength of an electron, from Eq. (5).

$$2\pi r = n\frac{h}{mv}, n = 1, 2, 3 \ldots \tag{7}$$

Finally we can rearrange Eq. (7) by multiplying both sides by mv and dividing both sides by 2π.

$$mvr = n\frac{h}{2\pi} = n\hbar, n = 1, 2, 3 \ldots \tag{8}$$

But we can see that Eq. (8) is identical to Eq. (4). We have derived Bohr's condition for stable orbits without ever having to make the special and unconnected assumption of quantization of angular momentum! The first Bohr orbit is simply the one in which one electron wave exactly fits, the second, one in which two electron waves fit, and so forth.

Louis de Broglie's contribution then did not banish the wave-particle duality but rather extended it to material particles. Yet in this extension there is a new unity. Light is not so very different from the rest of matter. In fact, in modern theories the primary difference between electrons and photons is that the former have a nonzero rest mass (that is, their mass when not moving) while the latter have a zero rest mass, which means that photons have no existence at rest but if present are always moving at the speed of light.

The apparently unavoidable fact is that light behaves like waves in its propagation through space and like particles in its interaction with matter. It has both wave and particle properties, and the more any experiment reveals one aspect, the less it reveals the other. Perhaps it should not be too surprising, after all, that we cannot flatly say that light is waves or light is particles. For when we say that something is made up of waves, we are really saying that it behaves like some large-scale motion that we have seen, such as that of water waves or waves on a stretched string. Similarly, when we say something is made up of particles, we are really saying that it behaves like thrown rocks, falling raindrops, or fired bullets. Waves and particles are concepts derived from the macroscopic (large-scale) world, and there is no compelling logical reason why light should behave like *anything* from the macroscopic world. Light is what it is, and if we can model it with only two concepts from the macroscopic world, maybe we should consider ourselves lucky.

QUESTIONS

1. What is a continuous spectrum and what produces it?
2. What is a line spectrum and what produces it?
3. Stars make fairly good blackbody radiators. If the sun is white in color and the giant star Beteleguse is reddish in color, which do you suppose has the higher surface temperature? Why?
4. Who was the founder of quantum mechanics? When?
5. Who first correctly explained the photoelectric effect?
6. Who discovered the atomic nucleus?
7. Who first used quantum theory to explain the nuclear atom and line spectra?
8. What is quantization of angular momentum and how was it used to explain the electron orbits of the hydrogen atom?
9. What is the energy of a quantum of ultraviolet radiation with a frequency of 10^{15} Hz?
10. If your body is emitting radiation with a wavelength of 9.4 μm, what is the energy of the photons?
11. What did A. H. Compton have to do with the quantum theory of light?
12. Who first postulated that material particles may have wave properties?
13. Find the wavelength of a baseball (m = 0.4 kg) thrown at 30 m/s.
14. Find the wavelength of an electron traveling at one-tenth the speed of light.
15. For an electron in the fourth Bohr orbit of hydrogen, how many electron waves are in the orbit? What is the value of the electron's angular momentum?
16. In what ways are electrons and photons alike? In what ways are they different?

Part II

Manipulation of Light

Chapter 4

Geometrical Optics

Fortunately there are a number of useful things we can do with light without having a detailed knowledge of its ultimate nature. A great deal of the theory of imaging does not depend at all on whether one regards light as waves or particles. In fact, much was known about imaging long before there was even any classical theory of light: the ancient Greeks and Romans had lenses or "burning glasses," and eyeglasses were known in Europe by 1300.[1] For this purpose, simple empirical rules governing the behavior of light rays will suffice. These rules for light rays comprise the field of **geometrical optics.**

4.1 LIGHT RAYS

A **light ray** may be thought of as a very thin, parallel beam of light, or it may be thought of as the path of a photon. Alternatively, it may be considered as an arrow showing the direction in which a wavefront moves. Most importantly, light rays show the direction of motion of light at any point, regardless of what one considers light to be. Although light rays are really hypothetical constructs, intended to help us trace light through optical systems, they can be approximated physically with thin beams or pencils of light, as just mentioned.

The most basic rule describing light rays is that in a single, unchanging medium they travel in *straight lines*. Indeed, we use light rays to define a straight line in prac-

[1] Ernst Mach, *The Principles of Physical Optics: An Historical and Philosophical Treatment,* trans. John S. Anderson and A. F. A. Young (New York: Dover Publications, Inc., 1926), pp. 50–51.

tice. In sighting down a pool cue or a 2 × 4 board to see if it is straight, a person is really comparing it with a light ray from the far end, assuming that the light ray is perfectly straight. When scientists speak of the rectilinear propagation of light (see Chapter 2), they are referring to this property of light rays.

Another basic rule for light rays is that they are *reversible.* That is, a ray will exactly retrace its path through an optical system in the opposite direction if the exiting ray is turned around and sent back along its path. In other words, when a light ray is drawn, the arrow can be placed on either end without violating the rules; it all depends on the direction in which it started.

4.1.1 Pinhole Images

Even the simple straight-line propagation rule by itself can explain some imaging effects. It may surprise you to learn that a device as simple as a small opening in an otherwise opaque screen can be used to form images. Such an opening is usually called a **pinhole** and the images formed are called **pinhole images.** We shall return to the formation of pinhole images, but first we need to consider what we mean by an image in general.

Consider an object every point of which is a source of light rays, either because the object is self-luminous or because it is reflecting light. If some of the light rays from one point A on the object are somehow brought together again at another point A' in space, that point A' is said to be the **image** of A (see Fig. 4–1). Since the rays from a single point tend to move apart, or diverge, from each other, you can see that some optical system is required to bring them together again. We may assume that if some optical system is acting to form an image of one point A, then it is doing the same for all the points of the object in a continuous fashion, so that the image of the whole object is a replica (perhaps enlarged or reduced) of the original object.

A pinhole forms an image, not exactly by bringing rays from a point back together, but rather by only allowing the passage of rays that never got very far apart (see Fig. 4–2). The only requirements for a pinhole to form an image are that it be small compared with the distance to the object and also with the distance to the viewing screen on which the image is. The former condition ensures that only a small, nearly parallel bundle of rays will be accepted from the object, while the latter condition makes the shape of the pinhole unimportant.

The advantages of a pinhole imaging system are that it has an infinite depth of field and infinite depth of focus. That means that all objects from very close to the pinhole on out are simultaneously in focus, and also that the viewing screen or image

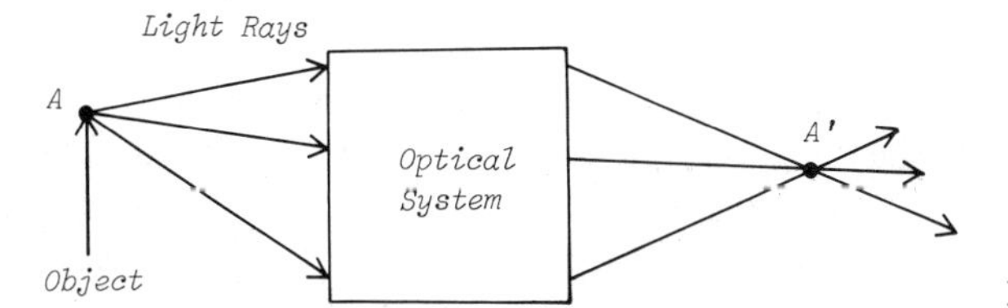

Figure 4–1 Image of a point source

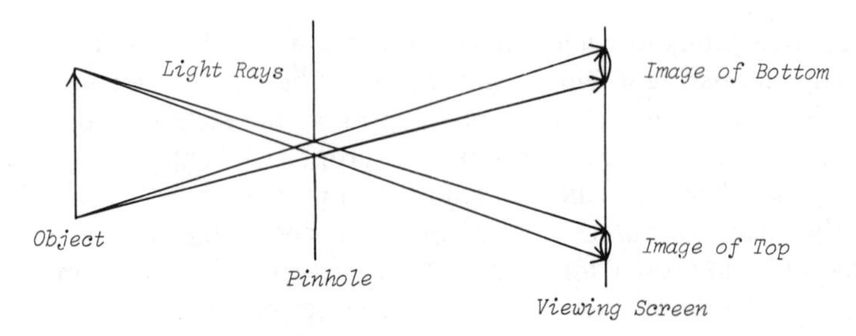

Figure 4–2 Pinhole image

plane can be put anywhere behind the pinhole (subject to the condition just mentioned). The major drawback to the pinhole image is that there is a direct trade-off between image *brightness* and image *sharpness* (called **resolution**). For a sharp image the pinhole should be very small (see Fig. 4–2), but a very small pinhole drastically limits the amount of light in the image, or its brightness. One can open up the pinhole for a brighter, fuzzier image or close it down for a dimmer, sharper image.

Pinhole viewers can be made from a box with a pinhole on one side and a viewing screen opposite, the so-called camera obscura known as early as 1519.[2] Furthermore, a photographic pinhole camera can be made just by replacing the viewing screen with photographic film. But you can see pinhole images in nature also. On a bright sunny day in the shade of a dense tree, faint elliptical spots of light can be seen on the ground. A white card held perpendicular to the direction of the sun and intercepting one of these spots will show it to be circular: a pinhole image of the sun. The pinholes in this case are simply the small openings between overlapping leaves. A more striking display is produced during a partial solar eclipse, when all the faint ellipses become crescents, demonstrating that these are really images of the sun independent of the shapes of the openings through the leaves.

4.2 REFLECTION AND REFRACTION

Of course, if all light rays ever did were to travel in straight lines outward from the source, geometrical optics would be a very simple and rather uninteresting field. The really interesting things happen when light rays encounter a change in the medium; then, in general, they change their direction. The change can be a gradual one, as when light travels through air of varying density, or it can be abrupt, as when light strikes the boundary between two different substances, such as air and glass. The latter situation is shown in Fig. 4–3.

Here you can see two effects taking place at the boundary. The incident ray is split into two rays: one bounces back into the first medium and is called the **reflected** ray, the other continues into the second medium in a changed direction and is called

[2] Ibid., p. 12.

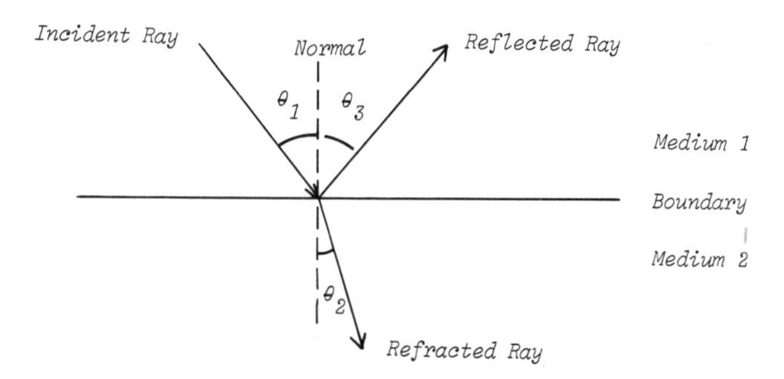

Figure 4–3 Reflection and refraction

the **refracted** ray. Angles are measured from the normal (perpendicular line) to the surface at the point of incidence. We have an angle of incidence θ_1 between the incident ray and the normal, an angle of reflection θ_3 between the reflected ray and the normal, and an angle of refraction θ_2 between the refracted ray and the normal. What fractions of the incident light appear in the refracted and reflected rays are determined by the nature of the two media and the angle of incidence; in general, the greater the angle of incidence, the greater is the proportion of reflected light. If you hold a piece of flat glass in daylight outdoors and view its surface at a very oblique angle it will act as almost a perfect mirror—you are seeing only reflected light. If you view the same glass surface perpendicularly you see light that has been refracted in passing through and little reflected light. For a glass-air interface and a perpendicular incident ray (0° angle of incidence), only about 4 percent of the light is reflected, while 96 percent goes into the refracted ray.

4.2.1 Reflection

It will be convenient to study the details of reflection first. In reflection a simple rule governs the ray behavior: The angle of reflection equals the angle of incidence. Referring to the symbols of Fig. 4–3 we may write

$$\theta_1 = \theta_3 \tag{9}$$

Even this modest increase in our knowledge allows us to analyze some common situations, such as imaging in a flat or plane mirror. Figure 4–4 shows a point source S as an object in front of a plane mirror M a distance p away. To find the image of S formed by the mirror we need to follow some light rays from S as they reflect from M. Actually, any two such rays from S will suffice, because the point at which they intersect after reflecting from M will establish the image. Since any two rays will do, it behooves us to choose ones that are simple to follow. One particularly simple ray is that which strikes M perpendicularly and is labeled 1 in the figure. The angle of incidence is zero, so the angle of reflection is also zero, which means that the ray

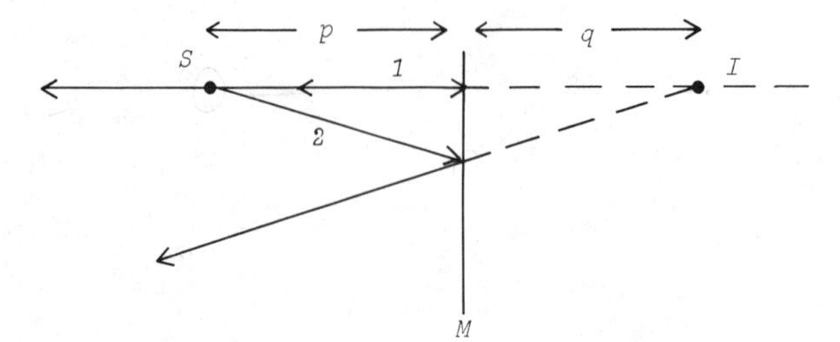

Figure 4–4 Image in a plane mirror

bounces right back on itself. A second, more general ray is labeled 2 in the diagram. You can see that the two rays after reflection diverge and will never intersect. Does that mean there is no image? Not really, because if an observer sees the two reflected rays, they both *appear* to have come from a single point behind the mirror, labeled I. This point, found by extending the reflected rays backward until the extensions intersect, is the image of S in the mirror. Using the law of reflection and some geometry, it is not difficult to prove that the image distance q must be equal to the object distance p; that is, the image is as far behind the mirror as the object is in front, a fact that should be familiar to you from viewing your own image in plane mirrors. It is important to remember that any ray from S that strikes M will reflect so that it appears to have come from I. This fact is illustrated in Fig. 4–5.

You may have noticed an important physical difference between the image in a plane mirror just discussed and the pinhole image studied earlier. The light rays do not really pass through the plane mirror image. The pinhole image could be seen on a viewing screen or even recorded on photographic film placed at the image position. Such actions are not possible with the plane mirror image because there is not really any light at the image position—the light only appears to have come from there. For these reasons we speak of two different kinds of images, real and virtual. A **real image** is one in which the light rays really come together in the points of the image. A **virtual image** is one in which the light rays only appear to have come together in the points of the image, but do not really do so. For practical purposes, a real image can always be seen on a viewing screen placed at the image position, while a virtual image cannot. Thus your familiar image in an ordinary mirror is a virtual image.

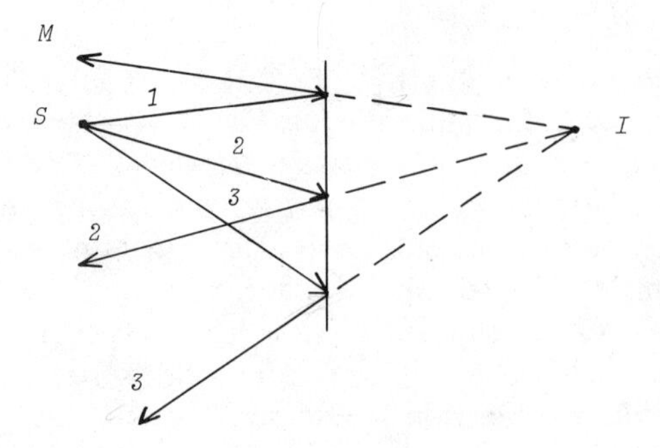

Figure 4–5 Plane mirror image rays

4.2.2 Refraction

The phenomenon of refraction is more complicated than reflection because the angle of refraction depends not only on the angle of incidence (see Fig. 4–3) but also on the nature of the two media involved. In particular, refraction depends on a quantity, defined for each medium, called index of refraction. The **index of refraction** for any medium is defined as the ratio of the speed of light c in a vacuum to the speed of light v in the medium. Using n for the index of refraction we have

$$n = \frac{c}{v} \tag{10}$$

Example:

What is the index of refraction of a plastic in which the speed of light is 2.37×10^8 m/s?

The speed of light in a vacuum is always $c = 3 \times 10^8$ m/s and we are given $v = 2.37 \times 10^8$ m/s.

Then $n = \dfrac{c}{v} = \dfrac{3 \times 10^8}{2.37 \times 10^8}$

$n = 1.27$

Example:

What is the speed of light in water if its index of refraction is 1.33? Here we know $n = 1.33$ and $c = 3 \times 10^8$ m/s.

$$n = \frac{c}{v}$$

$$v = \frac{c}{n} = \frac{3 \times 10^8}{1.33}$$

$$v = 2.26 \times 10^8 \text{ m/s}$$

There are two things to remember that will help you use Eq. (10). One is that the speed of light in a vacuum is always the same, $c = 3 \times 10^8$ m/s. The other is that c is the highest possible speed for light (or anything else), and it must be larger than v. Thus $n = c/v$ is always greater than 1. Additionally you might note that since the units of c and v are the same, they cancel out in the ratio and n has no units.

The index of refraction is a constant for a given material and is tabulated in many reference books (see Table 4–1). Diamond has one of the highest refractive indices of any substance. Since the refractive index is related inversely to the speed of

TABLE 4–1 REFRACTIVE INDICES OF COMMON MATERIALS

Medium	Index of refraction
Air (0°C, atmospheric pressure)	1.000292
Water	1.33
Ethyl alcohol	1.36
Glass	1.46–1.96
Quartz	1.54
Diamond	2.42

light in the medium, light must travel more slowly in diamond than in most other media. Actually the index of refraction can be slightly different for different wavelengths of light and also for different polarizations of light, but for the present we may ignore such variations. Roughly speaking, denser materials have higher indices of refraction.

The law of refraction can be stated in terms of the angles of incidence and refraction and the refractive indices of the two media involved. However, the mathematics needed is of higher level than we wish to use. Instead we shall consider some qualitative results of the law of refraction. Figure 4–3 illustrates the case in which medium 2 has a greater refractive index than medium 1. In this situation the ray bends toward the normal and the angle of refraction is less than the angle of incidence. However, light rays are reversible, and the arrows on the two rays could have pointed in the opposite direction. This situation is illustrated in Fig. 4–6. Here the ray is traveling from a medium of higher to one of lower refractive index and bends away from the normal, so that the angle of refraction θ_1 is greater than the angle of incidence θ_2. In general, when a ray travels from a medium of lower to one of higher refractive index, it bends toward the normal, and when it travels from one of higher to one of lower refractive index, it bends away from the normal. In any case, increasing the angle of incidence will increase the angle of refraction.

These facts of refraction become more understandable when we consider waves passing through the boundary at an angle and realize that a changing index of refraction simply means a changing wave speed. Let us redraw Fig. 4–3 to show the incident and refracted wave fronts (Fig. 4–7). Again, medium 2 is assumed to have a greater

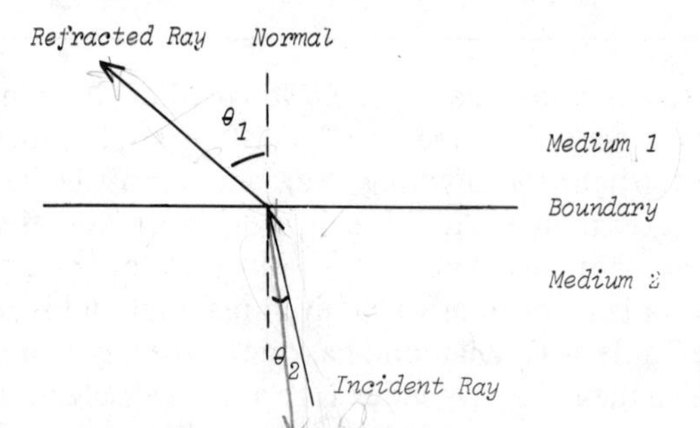

Figure 4–6 Refraction in a less dense medium

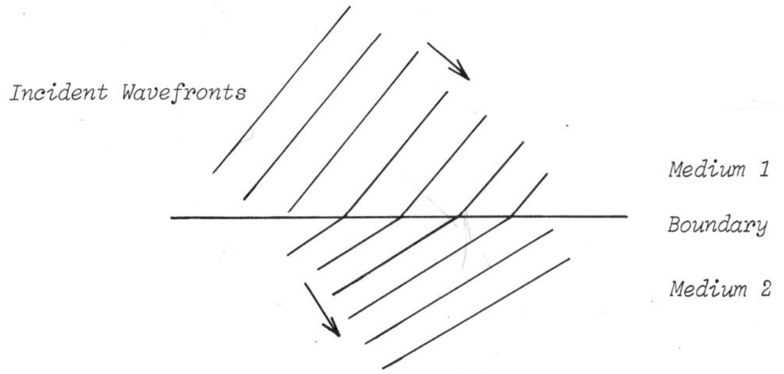

Figure 4–7 Refracted wave fronts

index of refraction than medium 1. But that greater refractive index implies a lower wave speed, according to Eq. (10). Thus in Fig. 4–7 the right-hand side of each wave front enters medium 2 first and slows down first, while the left-hand side continues at the higher speed typical of medium 1. This difference in speed turns the whole wave front to the right. In other words, the direction of the wave fronts bends toward the normal. It is very much like the movement of the ranks of a marching band as it wheels around to a different direction: one end of each rank must slow down while the other end keeps moving rapidly. Furthermore, it can be seen that the reverse effect (bending in the opposite direction) would occur if the wave fronts were turned around so that they proceeded from medium 2 to medium 1. Finally, it is easy to visualize what would happen if the light rays were perpendicular to the boundary (wave fronts parallel to the boundary): the light would change speed but no bending would occur because the whole wave front changes speed at once.

Earlier we said that the proportions of reflected and refracted light depend on the nature of the two media. Now it can be more precisely stated that they depend on the two refractive indices. The more nearly equal the two indices, the less light is reflected. If the two media have the same index of refraction, then no light is reflected, the angle of refraction equals the angle of incidence, and as far as the light is concerned, there is no boundary.

One effect of refraction can be seen when you look down into a clear pool of water. The pool always appears shallower than it really is. Figure 4–8 illustrates this effect by tracing two rays from a point on the bottom labeled *O*. One of the rays used

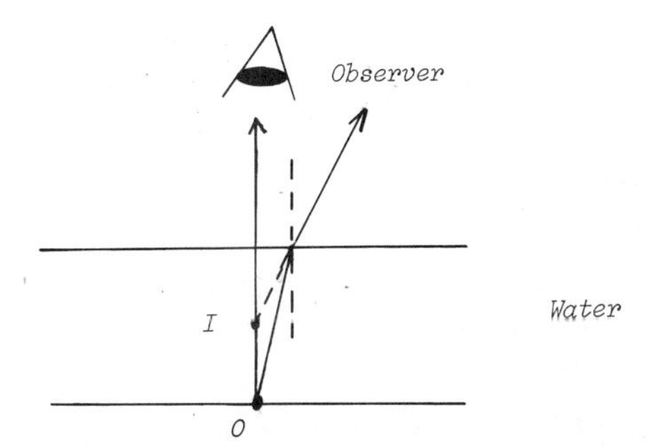

Figure 4–8 Depth illusion

has a vertically upward direction, striking the water surface perpendicularly and therefore not bending. The other ray is shown to the right, making a small angle of incidence at the surface and bending away from the normal as it enters the air. To an observer looking down into the water and seeing both these rays they appear to have come from the point labeled *I* rather than from the point *O*. Therefore the bottom appears to be at the shallower depth of *I*. This effect is even more pronounced if the observer is looking in at an angle rather than straight down. Then the observed rays must have larger angles of incidence and refraction at the surface, causing an even greater illusion, as shown in Fig. 4-9.

You can observe the whole range of effects with a flat, clear glass or plastic paperweight that has writing or a design facing upward on the bottom. As you look in from the top and tilt the paperweight so you are viewing it more and more obliquely, you can see the bottom design apparently move closer and closer to the top. A limiting case is reached when you are looking in almost parallel to the top surface. Then you are seeing rays just skimming the surface, with an angle of refraction of nearly 90°, and the bottom surface appears to be right at the top. Figure 4-10 shows this case.

4.2.3 Total Internal Reflection

The situation illustrated by Fig. 4-10 serves to introduce us to another interesting aspect of refraction. When light travels from a medium of higher refractive index to a medium of lower refractive index, as in the case above, the rays bend away from the normal so that the angle of refraction is greater than the angle of incidence. As the angle of incidence increases, the angle of refraction also increases until eventually it becomes 90°; the refracted ray will then be just skimming the surface. At this point the angle of incidence will still be less than 90°. What happens if we now increase the angle of incidence still further? The angle of refraction cannot increase and still give a ray in the second medium. Instead, the refracted ray disappears altogether and all the light is reflected at the surface. This phenomenon is known as **total internal reflection.** The angle of incidence when the angle of refraction is 90° is known as the critical angle θ_c, and any angle of incidence greater than the critical angle produces total internal reflection, as shown in Fig. 4-11.

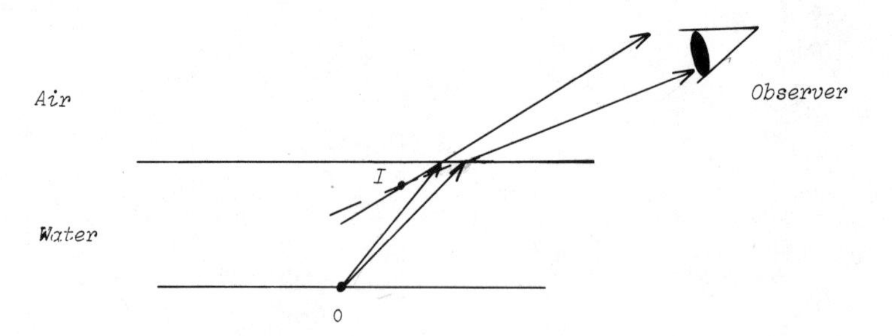

Figure 4-9 Depth illusion at oblique angles

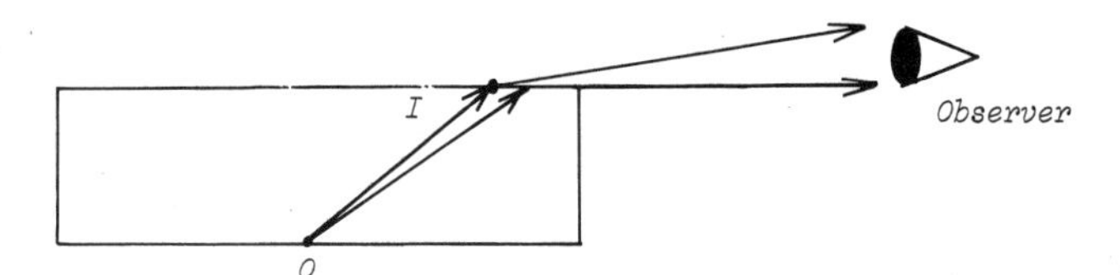

Figure 4–10 Angle of refraction of 90°

It is important to remember that total internal reflection can only occur when light is traveling from a medium of higher refractive index to one of lower refractive index, as from glass to air or water to air. If light is traveling the other way, the angle of incidence is always greater than the angle of refraction, so the latter angle can never reach 90°. In any case, the fraction of the light that appears in the reflected ray increases with increasing angle of incidence, but only in total internal reflection does that fraction become 100 percent for incident angles less than 90°.

The critical angle depends on the refractive indices on the two sides of the boundary. Usually the medium with lower refractive index is a vacuum ($n = 1$) or air ($n \cong 1$). With the lower refractive index fixed at unity, the critical angle then depends only on the index of refraction of the remaining medium. The greater this index is, the smaller the critical angle under these assumptions. Thus water, with $n = 1.33$, has $\theta_c = 45.8°$, glass, with $n = 1.5$, has $\theta_c = 41.8°$, and diamond, with $n = 2.4$, has $\theta_c = 24.6°$. Remember that a smaller critical angle implies a greater likelihood of total internal reflection because the angle of incidence is more likely to be greater than θ_c in that case.

Diamond has one of the highest known indices of refraction and therefore one of the smallest critical angles in air of any material. This small critical angle means that light inside diamond is more likely to be totally internally reflected (any angle of incidence greater than 24.6° gives the effect). This effect accounts for the brilliant sparkle of a diamond when cut as a gemstone. Light that enters the diamond at one facet is usually totally internally reflected several times, without any loss in intensity, before exiting from another facet in another direction. Thus one sees unexpected flashes of light when looking in. Of course the various diamond cuts are designed to enhance this effect. Figure 4–12 shows some possible ray paths through a diamond cut as a so-called brilliant. The colors seen in a diamond come from an effect to be explained later.

Total internal reflection is also used in several kinds of glass prisms for reflect-

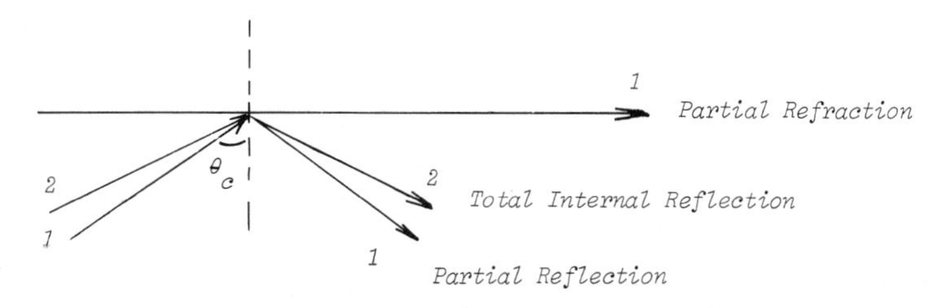

Figure 4–11 Total internal reflection

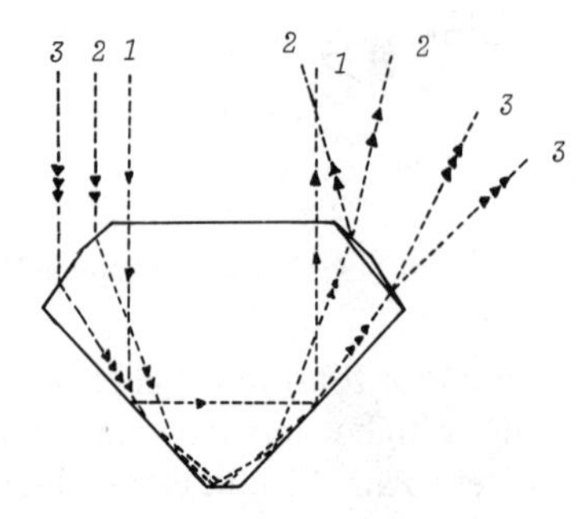

Figure 4–12 Ray paths in diamond

ing light. Most glass has an index of refraction of 1.5 or more, giving a critical angle less than 45°. Therefore an angle of incidence of 45° inside the glass will produce total internal reflection. This fact can be put to use in a simple glass triangular prism with angles of 45°, 45°, and 90°, as shown in Fig. 4–13.

The first case shows the prism used to bend light 90°. The ray enters one side of the prism perpendicularly, strikes the hypotenuse at an angle of 45°, totally internally reflects at an angle of 45°, and exits perpendicularly to the other side. The second instance shows the prism producing a 180° turn. Here the ray enters perpendicularly to the hypotenuse, strikes one side at an angle of 45°, totally internally reflects at an angle of 45°, repeats the same sort of reflection at the other side, and finally exits perpendicularly to the hypotenuse again. It is interesting to note that you still get the 180° reversal of the light even if it is not quite perpendicular to the hypotenuse when it enters. However the picture shows only a two-dimensional cross section so that only reversal of the direction of the light in the plane of this cross section can be shown. In a three-dimensional view the ray could be coming in below the plane of the paper and leaving above it. To reflect a beam of light back on itself in three dimensions, one needs three perpendicular faces meeting at a corner, as shown in Fig. 4–14. Such a prism is called a corner reflector and serves to turn any beam of light entering at any angle right back on itself. An array of 100 such corner reflectors was left on the moon by the first astronauts so that scientists could send a laser beam through a telescope on the earth to the reflector on the moon and receive it back in the same telescope. The time of flight of such a light pulse provides an accurate measurement of the distance between the earth and the moon.

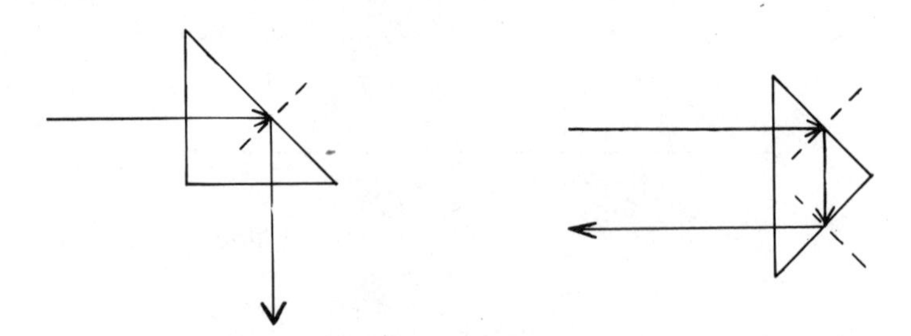

Figure 4–13 Glass 45° prism

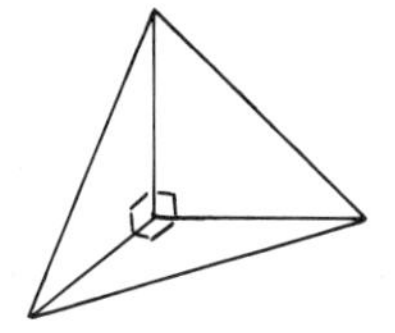

Figure 4–14 Corner reflector

One result of total internal reflection you may enjoy verifying yourself, especially if you have your own swimming pool so that you can obtain the flat, quiet surface needed. When you are underwater, looking up you should be able to see all around the horizon (or at least the edge of the pool) in a cone above your head with a half-angle of 45.8°, the critical angle of water. Outside of that cone on the surface you should be able to see reflections of the bottom of the pool. The situation is shown in Fig. 4–15. This figure is really identical to Fig. 4–11, but with the rays reversed in direction. Since light rays are reversible, Fig. 4–15 must be a correct picture.

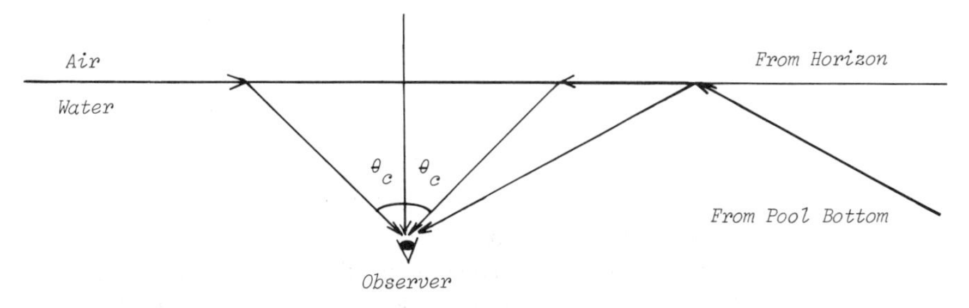

Figure 4–15 Underwater view

4.3 SPHERICAL LENSES AND MIRRORS

Thus far our discussion has been exclusively about flat or plane surfaces as the boundaries beween media. When the surfaces are curved instead, then focusing becomes possible. That is, reflection or refraction at a curved surface can cause rays that were originally parallel to each other to become nonparallel and therefore to intersect. The most commonly used curved surface for focusing is a section of a sphere. A spherical mirror focuses by reflection, while a spherical lens focuses by refraction, as shown in Fig. 4–16. Here, in both cases a parallel bundle of rays is shown striking the spherical surface and being changed to a converging bundle of rays intersecting at a point called the **focal point.** The original ray direction is along the **axis,** which is a line drawn perpendicular to the mirror or lens at its center; and the focal point is also on the axis. No new physical laws are involved here; the focusing is a result of the law of reflection or refraction and of geometry. The spherical surfaces can be viewed on a microscopic scale as being composed of small planes oriented in different directions,

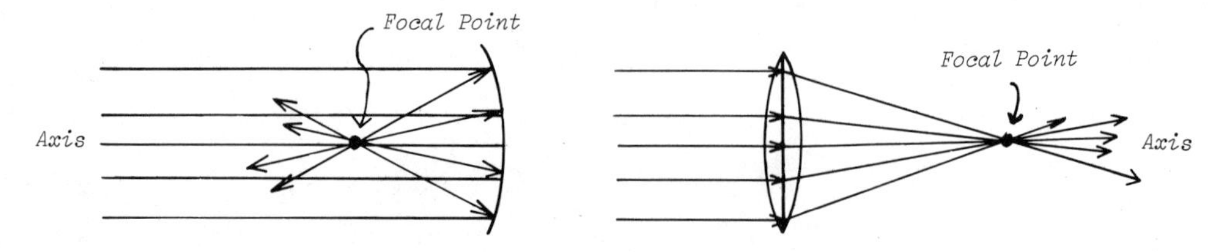

Figure 4–16 Spherical mirror and lens

and then rays and normal lines can be treated for each plane just as was done earlier. In the lens, refraction takes place at each of the two spherical surfaces, but we usually draw just one ray bending in the middle of the lens to show the total effect. You should also keep in mind that the drawings of Fig. 4–16 are cross sections of three-dimensional objects, which are shown in perspective in Fig. 4–17.

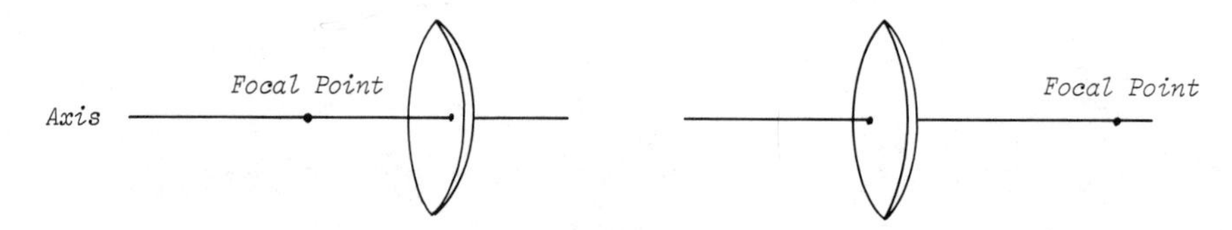

Figure 4–17 Spherical mirror and lens

You can see from Fig. 4–16 that the focusing effect is at least plausible in terms of light rays and the rules that govern their reflection and refraction. This focusing is also reasonable from the point of view of wave fronts. Figure 4–18 shows plane waves, which correspond to parallel light rays, incident on the spherical lens and mirror, which transform them to spherical waves centered on the focal points. When the plane waves strike the mirror, the outer edges strike first and reflect. Only later does the center of the wave reflect, so that the reflected wave front is no longer flat but has leading outer edges and a trailing center, that is, it forms spherical waves converging to the focal point. When the plane waves strike the lens, the centers move through a thicker portion of the lens than do the edges. Therefore portions of the waves nearer the axis are slowed down longer and emerge on the other side farther behind, spherical waves converging to the focal point being again formed.

A spherical lens or mirror has a similar focusing effect on parallel rays even if

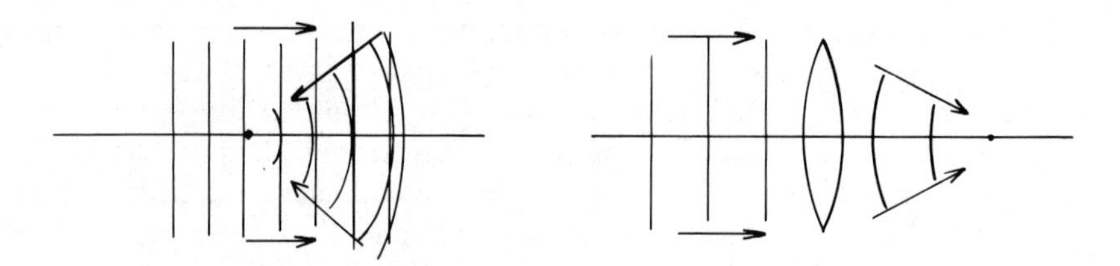

Figure 4–18 Wave fronts with mirror and lens

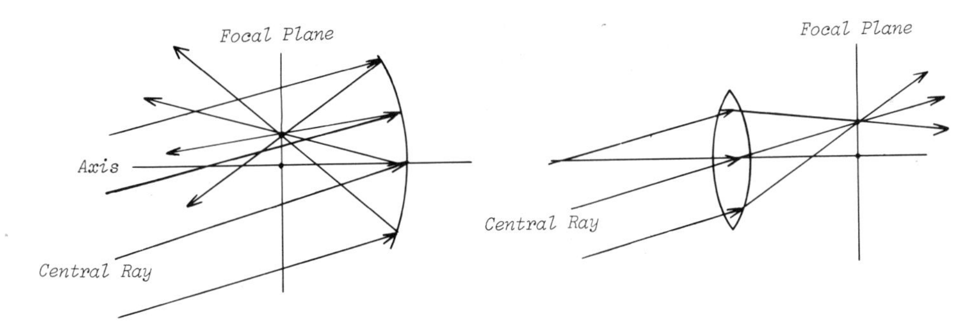

Figure 4–19 Focal plane

they are incident at some angle to the axis. Then the rays are brought to a focus at an off-axis point located at the same distance from the mirror or lens as its focal point. For this reason we speak of a **focal plane** at a distance from the lens or mirror called the **focal length** *f*. Figure 4–19 illustrates these concepts. The point in the focal plane to which the rays converge can be found by where the central ray (striking the center of the lens or mirror) crosses the focal plane. For the mirror the central ray just reflects at an equal angle on the other side of the axis. The ray through the center of the lens continues undeviated, because at the center the two sides of the lens are parallel and parallel faces of a material in air will not change the original direction of a ray. Each spherical lens and mirror has a focal length which depends on the curvature of the surfaces and, in the case of the lens, on the index of refraction. The mathematical formulas for focal length are given in Appendix A.

Actually there is a type of spherical mirror and a type of spherical lens other than those shown in Figs. 4–16 and 4–17. The mirror surface could be convex and the lens surfaces could be concave, as shown in Fig. 4–20. Here the rays, originally parallel, diverge after striking the optical device rather than converging. Therefore they never intersect. But we can still define a focal point: we use the point the rays appear to have come from, just as we did to identify the image point for a plane mirror, by extending the reflected or refracted rays backward. In this case we may speak of a

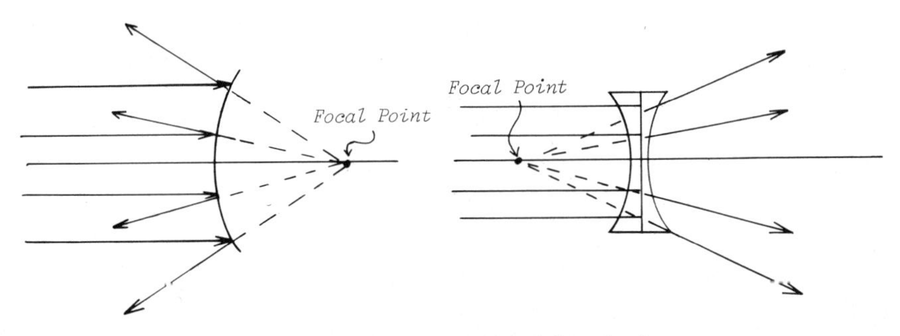

Figure 4–20 Convex mirror and diverging lens

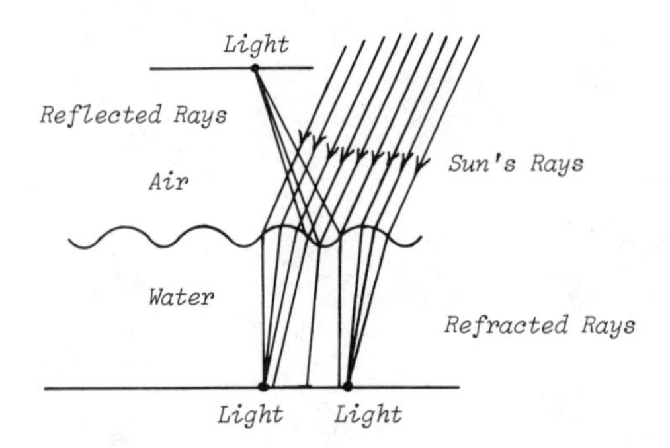

Figure 4–21 Focusing by water waves

virtual focal point. The lens of Fig. 4–20 is called a **diverging lens,** as opposed to the **converging lens** of Fig. 4–16.

One beautiful result of the focusing of light by curved surfaces in nature may be seen when sunlight falls on the gently undulating surface of a pool of water. The alternating convex and concave water surface acts to focus the light by refraction or reflection, so that rippling lines of light can be seen on the bottom of the pool or on a surface above the pool that catches the reflection. The lines of light move along with the water waves and show the same random irregularities. Figure 4–21 illustrates this effect.

In reality, the focusing properties of spherical lenses and mirrors are not perfect. Parallel rays are not brought together precisely at a single point but only approximately so. The rays incident near the edge of the lens or mirror intersect slightly closer to the lens or mirror than those incident near the center. This effect, known as **spherical aberration,** is shown exaggerated in Fig. 4–22. To keep spherical aberration small and unimportant we must keep the size of the lens or mirror small compared with its focal length.

4.3.1 Imaging

Of course, lenses and mirrors are not just used for focusing parallel beams of light but for forming images. In a way, the focusing of a parallel beam of light can also be considered to be the formation of an image. The parallel rays could all be thought of as coming from a single point object infinitely far away. Then where they are brought together again, the focal point, would be the image of that distant object. You can see from Fig. 4–23 that more distant point objects produce more nearly parallel rays

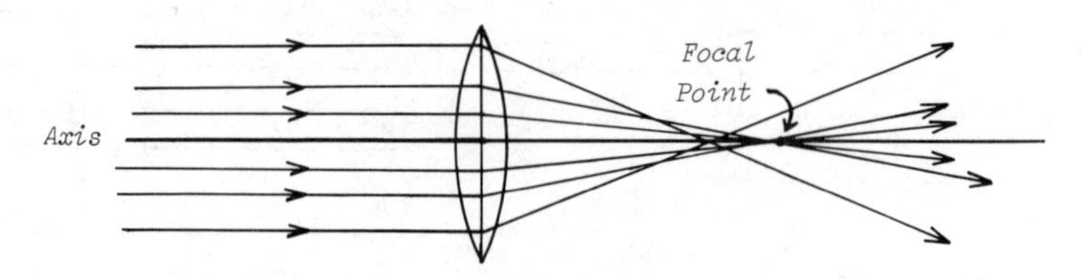

Figure 4–22 Spherical aberration

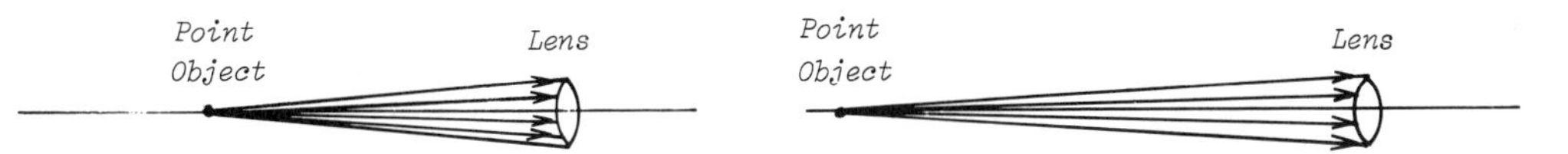

Figure 4–23 Nearly parallel rays from distant object

striking a lens. An actual object can never be infinitely far away, but all that is required for this viewpoint to be valid is that the object be much farther away than the focal length. A point object infinitely far away on the axis of a spherical lens or mirror produces an image at the focal point. A point object infinitely far away off-axis produces an off-axis image in the focal plane.

We can use the facts we already have about light rays, lenses, and mirrors to investigate imaging of objects closer than infinity. First we need to note one more property of lenses. A lens will always have two focal points, one on either side. If the lens is thin, the focal points are equally distant from the lens and the lens can be used when turned either way. We can say the lens has the same focal length from either side. A thick lens still has the same focal length on either side, but it is not immediately obvious from where this focal length should be measured, as the two focal points are not necessarily equidistant from the lens center. All this discussion assumes that the same medium is present on both sides of the lens. From now on we will consider only the simpler case of thin lenses.

Suppose an extended object is placed near a lens but outside its focal point, as shown in Fig. 4–24. Three rays are traced from the top point of the object through the lens. Ray 1 starts out parallel to the axis and therefore is always refracted so that it passes through the focal point on the other side. Ray 2 passes through the center of the lens and so remains undeviated. Ray 3 passes through the focal point on the near side and is refracted by the lens to emerge parallel to the axis; this ray path demonstrates the reversibility of light rays because it is obviously correct if the ray is turned around. The point at which these three rays intersect on the other side of the lens is the image of the top point of the object. Any two of these rays would have determined this image point, and any ray from the object point which passes through the lens should bend so that it passes through the image point. Three similar rays from the bottom point of the object are traced through in Fig. 4–24 to also determine the image of that point. Tracing through rays from every object point would trace out

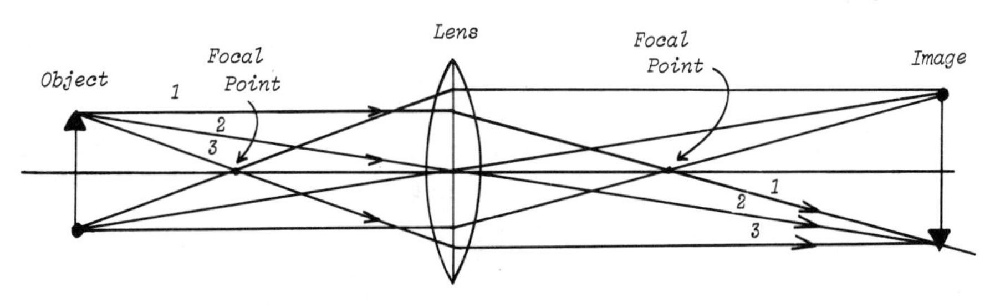

Figure 4–24 Imaging by a lens

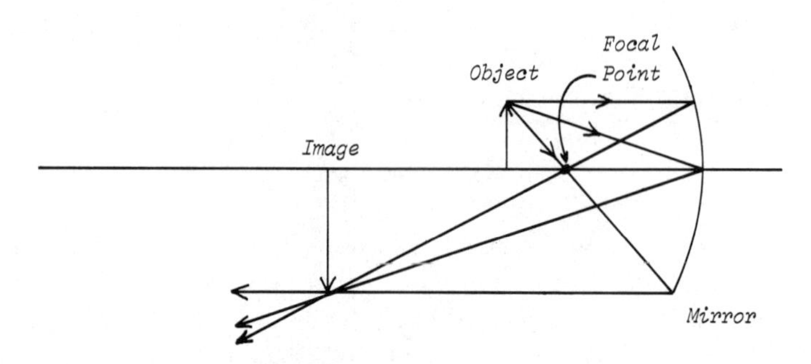

Figure 4–25 Imaging by a mirror

the full image shown. This image can be seen to be *real, inverted,* and *enlarged.* This is simply the case of a projection lens, such as that in a slide projector. In such a projector the slides are put in upside down and close to the lens and the image is upright, far away from the lens, enlarged, and real (it is shown on a screen).

A concave mirror can form a similar image in a similar manner, as shown in Fig. 4–25, where only the top halves of the object and image are shown for simplicity.

For either the converging lens or the concave mirror, the image distance and size depend on the object distance. If the object is close to the focal point (but still outside it), then the image is very far away and large. As the object is moved away from the focal point, the image moves in closer and becomes smaller. As we saw earlier, a very distant object produces an image essentially at the focal point. When the object is at a distance equal to twice the focal length, then the image is *at the same distance* and is the *same size.* Figure 4–26 illustrates this case. In fact, we can say that the real image and object are always interchangeable because light rays are reversible. For any object and image position (say Fig. 4–24, 4–25, or 4–26), we could turn the light rays around by simply reversing the arrows on them without otherwise changing their paths. But such a reversal makes the image into the object and vice versa, and we have another legitimate ray trace.

It is also possible to produce *virtual* images with a converging lens or concave mirror by moving the object inside the focal point. Figure 4–27 shows this result for a converging lens. The rays traced through the lens from a point on the object do not

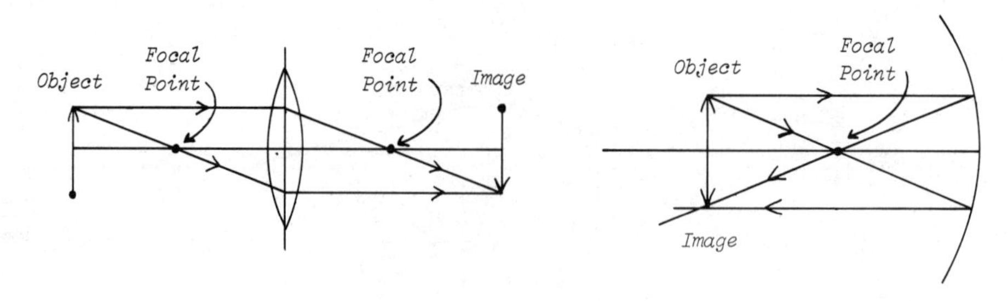

Figure 4–26 Equal object and image distance

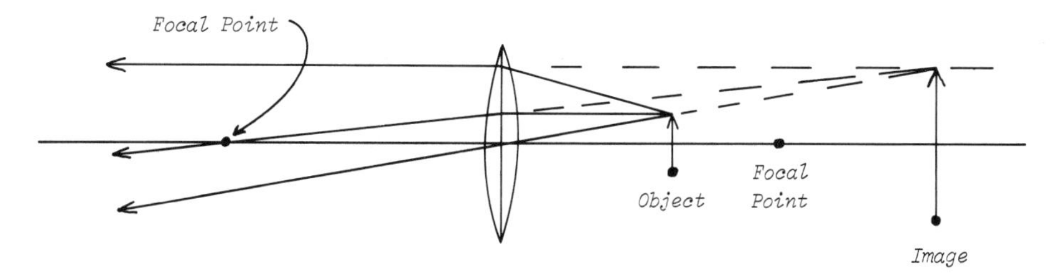

Figure 4–27 Virtual image from converging lens

intersect anywhere after refraction, but they do appear to have come from a point on the same side of the lens as the object. Therefore we can say there is a virtual image which can be found by extending the refracted rays backward until they intersect. You can see from the figure that the image is *virtual, upright,* and *enlarged.* This situation is just the case of a simple magnifier or magnifying glass. To use such an instrument you bring it up close to the object (placing the object inside the focal point) and look through it to see an upright, enlarged image, which appears to be on the same side of the lens as the object.

As an interesting little experiment, you can produce your own simple magnifier from a teardrop. First you must produce some tears, say by a forced yawn. Then you must look directly downward at something so that the teardrop can collect into a spherical shape directly over your pupil where the light enters your eye. You must not blink, for that clears the tears away. In fact it may be helpful to hold your eyelids wide open with thumb and forefinger to keep from disturbing the gathering teardrops. The object you are looking at will blur as the teardrop forms and when it does, move the object closer. Eventually, you will end up with the object almost touching your nose and with a highly magnified view. Figure 4–28 illustrates. This situation is somewhat different than we have discussed because there is not the same medium on each side of the teardrop lens, but the basic principles are the same. As one author has said, "We have this magnifier always with us, and can get a closer look at anything we can look down on, and cry over."[3]

The production of a virtual image by placing an object inside the focal point of a concave mirror is shown in Figure 4–29. The arrangement of Figure 4–29 is that of a shaving mirror. Such a mirror is slightly concave with a long focal length. You put

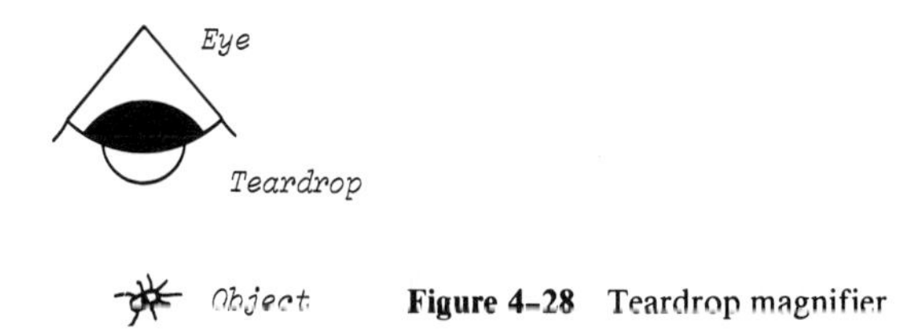

Figure 4–28 Teardrop magnifier

[3] Arthur D. Bates, *J. Opt. Soc. Am.* 58 (1968), 1018.

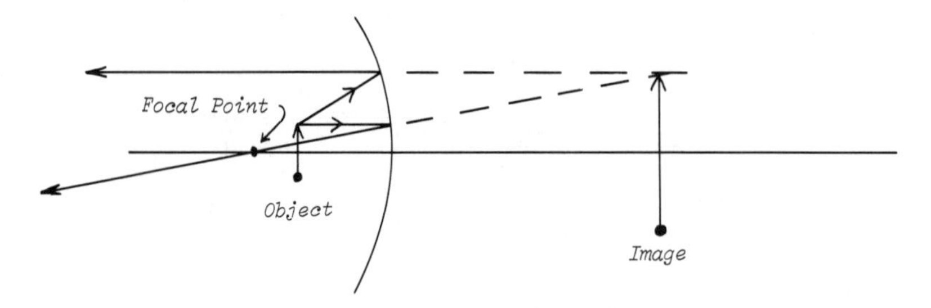

Figure 4–29 Virtual image from concave mirror

your face inside the focal point and obtain a *virtual, upright, enlarged* image behind the mirror.

For either a converging lens or a concave mirror, an object near the focal point (but inside it) will produce a very distant, enlarged, virtual image. As the object is moved closer to the lens or mirror, the image also moves closer and gets smaller, until at zero distance the object and image merge to the same position and size. Of course the real and virtual images for a mirror are on opposite sides from those for a lens. We can now see the full range of possible object and image positions for a converging lens or concave mirror. Starting with an object at infinity, we obtain a real image at the focal point. As the object is moved toward the optical device, the image moves away until the object reaches the focal point and the image reaches infinity. Now as the object passes through the focal point, the image seems to pass through infinity, changing from real to virtual and switching to the other end of the axis. It is almost as if the left end of the axis were connected to the right end of the axis at infinity, except that the image cannot pass through without a change in character. Finally, as the object continues approaching the optical device, the virtual image comes all the way in from infinity to the device, but from a direction opposite to the way it went out.

We shall just briefly mention here that a diverging lens or convex mirror always forms a *virtual, upright* image of a real object at any object distance. Figure 4–30 illustrates. You may have seen this type of mirror as a rearview mirror or a store security mirror. Sometimes called a wide-angle mirror, it always gives a reduced image to provide a larger field of view.

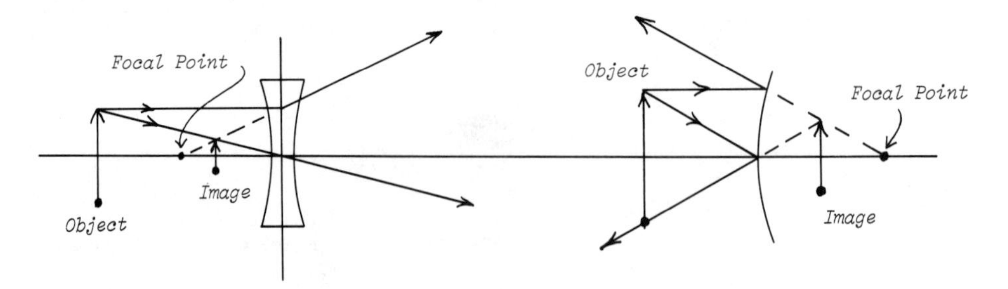

Figure 4–30 Imaging with diverging lens or convex mirror

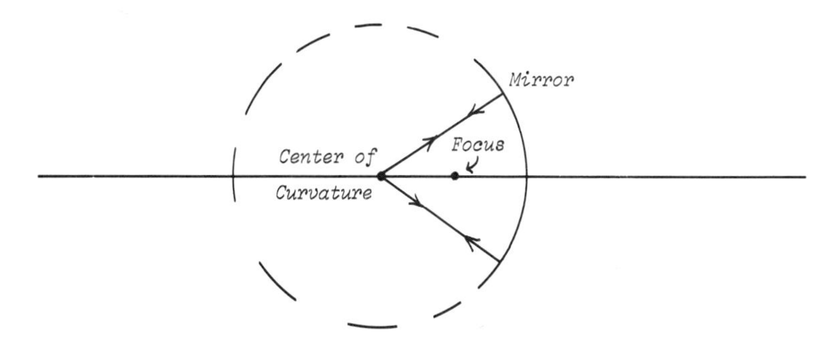

Figure 4–31 Center of curvature

All the situations we have considered with spherical lenses and mirrors can be analyzed mathematically. Relatively simple formulas give the image position, size, and character from the focal length and object position. The interested reader is referred to Appendix A for the mathematical treatment of spherical lenses and mirrors.

Because spherical lenses and mirrors do not focus perfectly (owing to spherical aberration), they also do not image perfectly. Points that are off-axis are subject to additional aberrations, but even points on the axis will be imaged with spherical aberration, that is, imperfectly. If we consider the simpler case of reflection from spherical mirrors, it is not too difficult to show that the only on-axis point that is perfectly imaged (neglecting wave effects, as is always assumed in geometrical optics) is the center of curvature of the mirror. This object point happens to be at twice the focal length of the mirror. If a point is the center of curvature of the mirror, it means that the mirror is a section of a sphere whose center is at that point, as shown in Fig. 4–31. Light rays drawn from an object at the center of curvature will strike the mirror perpendicularly, because a line from the center of a sphere to its circumference (a radius line) always meets the circumference at a right angle. Therefore every ray will have an angle of incidence of zero, and, by the law of reflection, an angle of reflection of zero, that is, the rays double right back on themselves. This ray behavior puts the image right back on the object at the center of curvature. Earlier we noted that an object at a distance of $2f$ produces an image the same distance away from the mirror, but the point here is that *all* the rays go *precisely* through the image no matter how large the mirror is—there is no spherical aberration.

4.4 CONIC SECTION MIRRORS

Spherical surfaces are not the only curved surfaces we could use to reflect light. Are there other curved surfaces which might give perfect imaging for other on-axis points? The answer is yes, and to obtain such results we need to consider a more general class of curves called conic sections, of which circles are only one type. The most

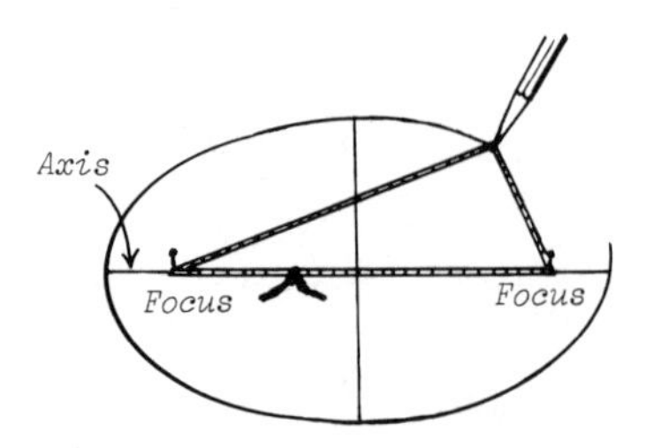

Figure 4–32 Drawing an ellipse

general example of a conic section is an ellipse, which you are probably familiar with as a kind of flattened circle. You can draw an ellipse by the following procedure: (1) put two thumbtacks through a sheet of paper securing it firmly to some backing; (2) put a loop of string over the thumbtacks so they are inside the loop but with slack left over; (3) put your pencil point inside the loop on the paper so that it takes up the slack and the string forms a triangle; (4) now move your pencil smoothly around the inside of the string keeping it taut and you will draw an ellipse. Figure 4–32 shows how. The points where the thumbtacks were are called the **foci** (each one a **focus**) of the ellipse, and a line through the foci is the **major axis** of the ellipse. Suppose we think of keeping the right-hand focus fixed but moving the left-hand one along the axis. You can probably see from the construction of Fig. 4–32 that as the left focus moves away from the other, flatter ellipses are produced, and that as it moves toward the other, rounder ellipses result. Figure 4–33 shows some examples. Furthermore, if one focus is moved right on top of the other, the ellipse becomes a circle with both foci at the center; the string in Fig. 4–32 becomes a single length line from the thumbtack to the pencil. We said earlier that an ellipse is a kind of flattened circle, but we should more properly call a circle a rounded ellipse because the circle is the special case of an ellipse with zero distance between foci.

The circle and the ellipse are two curves of the general class called **conic sections,** but there are also two others. To see them you have to stretch the ellipse and your imagination a bit more. Let us go back to the ellipse with a fixed right and moving left focus. This time imagine the left-hand focus moving away forever, or, to put it another way, displaced to infinity. Then the resulting ellipse is deformed into a curve open at one end, called a **parabola.** This curve is shown in Fig. 4–34. Like the circle, the parabola may be thought of as a special case of the ellipse, one with one focus displaced to infinity.

For the final curve you need to let your imagination pass through infinity with

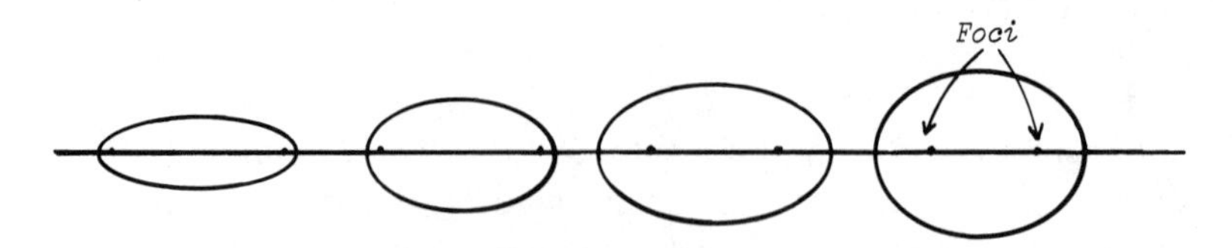

Figure 4–33 Ellipses

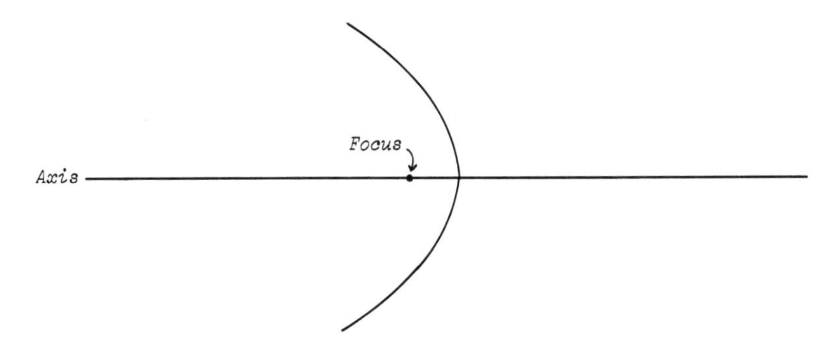

Figure 4–34 Parabola

the moving focus. As was the case with images for a converging lens, we can think of the focus as passing through infinity and coming back in from the opposite direction, as if infinity at the left hand end of the axis were somehow connected to infinity at the right hand end. Then the moving focus appears to the right with the other open-ended branch of a curve called a **hyperbola,** as shown in Fig. 4–35. To stretch a point (not to mention the ellipse), we can think of a hyperbola as an ellipse turned inside out through infinity.

The circle, ellipse, parabola, and hyperbola are called conic sections because they are produced at the intersection when a plane slices a right circular cone at various angles, as shown in Fig. 4–36. Slicing the cone at a right angle to its axis produces a circle. Slicing at a more oblique angle produces an ellipse. Slicing parallel to one side produces a parabola. Finally, a slice at a steeper tilt, which also cuts the extension of the cone on the other side of its apex, produces a hyperbola. If you have an ice cream cone, this result may actually be demonstrated, although only one branch of the hyperbola is obtained.

The conic sections can also be produced by the shadow of a ball on a flat surface. In this case, not only can you see the curve, but by tracing the shadow you can draw it. A spherical ball should be placed on a tabletop in a darkened room with a point source of light. The light held vertically over the ball produces a circular shadow centered over the point of contact between the ball and the table. The foci are at the contact point. If the light source is moved slightly off to one side, the shadow becomes an ellipse with one focus still at the contact point. The farther the light

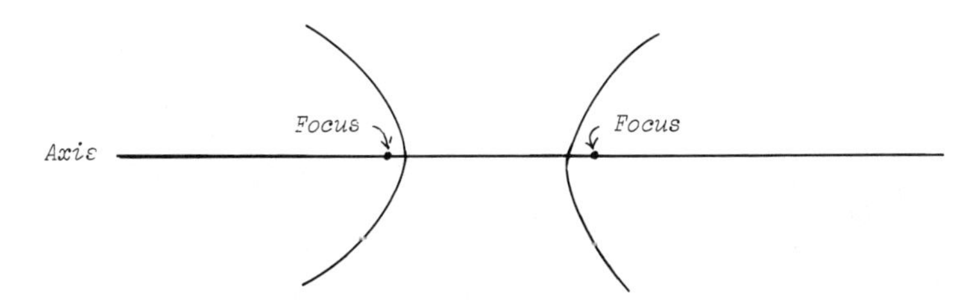

Figure 4–35 Hyperbola

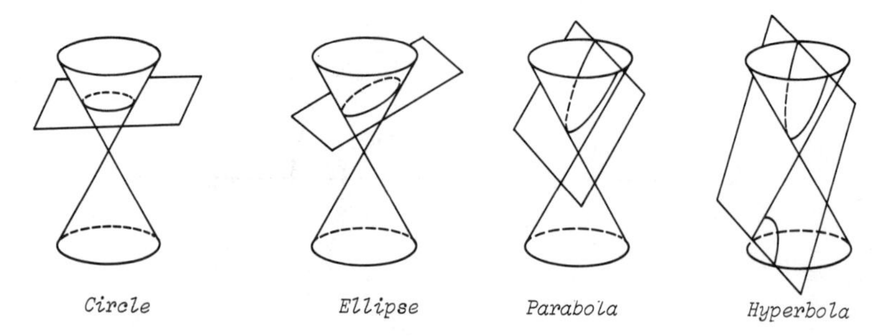

Figure 4–36 Conic sections

source is moved off vertical, the farther away the other focus moves and the more elongated the ellipse becomes. If the light source is put level with the top of the ball, the shadow becomes the open-ended parabola: one focus has moved to infinity while the other has stayed under the point of contact. These cases are illustrated in Fig. 4–37. Dropping the source of light below the top of the ball produces a shadow that is a hyperbola. The two branches of the hyperbola can even be produced if an identical second ball is held exactly opposite the first ball, the same distance away; then its shadow is the other branch, as shown in Fig. 4–38. This "counterball" could be considered to be present on the other side of the light in each of the cases, but its shadow will not fall on the table until the light is brought below the first ball. That is, only the hyperbola has two branches. These shadow conic sections should not be surprising because the shadow region of the sphere is a cone, which is being sliced by the plane of the table. However, in this visualization one of the foci is always easily identifiable as the contact point between the ball and the table.

To get back to imaging, the law of reflection and the geometry of ellipses assure us that any ray from one focus of an ellipse will be reflected so that it goes through the other focus. Each focus is a real image of the other in an elliptical reflector, precisely and without aberrations in geometrical optics. Figure 4–39 illustrates. Again we must be aware that Fig. 4–39 shows only a cross section of the type of reflector we are talking about. In three dimensions we have to imagine rotating an ellipse about its major

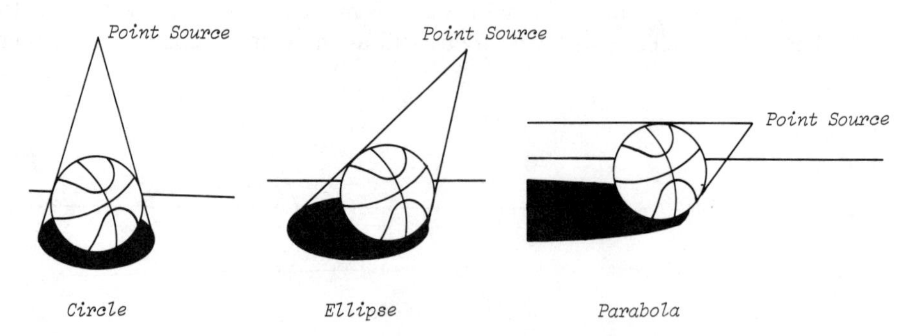

Figure 4–37 Circular, elliptical, and parabolic shadows

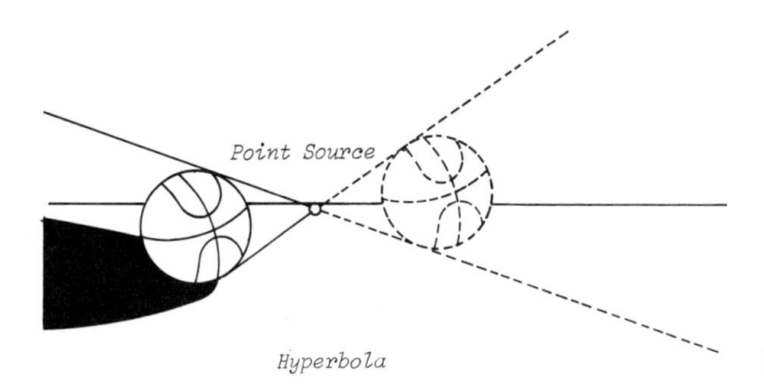

Figure 4–38 Hyperbolic shadow

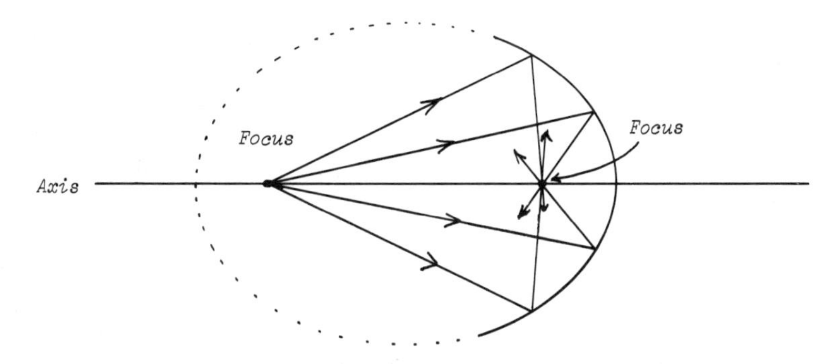

Figure 4–39 Elliptical mirror

axis to produce a figure called an **ellipsoid of revolution,** a section of which becomes our mirror as shown in Fig. 4–40. We have noted that the other conic sections can be considered as special cases of the ellipse. This viewpoint confirms our earlier analysis of a spherical mirror (circular cross section), which showed that a point object at the center was perfectly imaged right back at the center. When the circle is viewed as a special case of the ellipse, it must be considered to have two foci superimposed at its center. Then a point object at one focus, the center, must be perfectly imaged at the

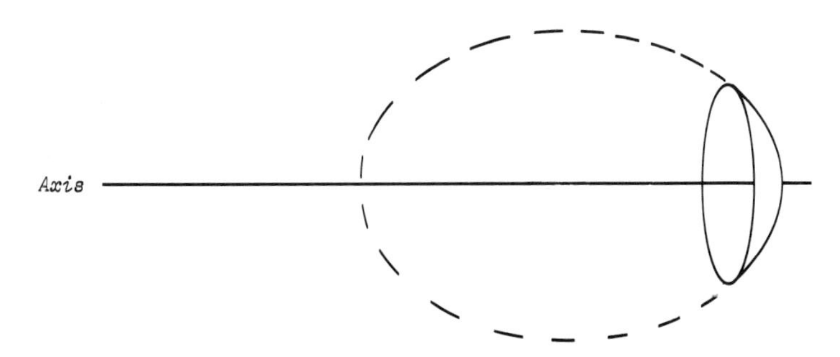

Figure 4–40 Ellipsoid of revolution

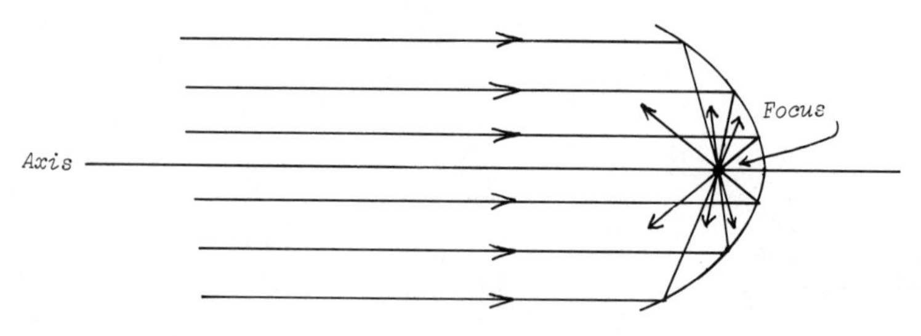

Figure 4–41 Parabolic mirror

other focus, also the center. Thus the two different approaches lead to the same result.

A **parabolic reflector** (more properly, a **paraboloid of revolution**) has one focus near its concave side and one focus at infinity. Therefore a point source at the near focus should be imaged at infinity and vice versa. But that is the same thing as saying that parallel rays along the axis will be reflected to a perfect point at the near focus or that a point source at the near focus will produce a bundle of rays all parallel to the axis. This result is shown in Fig. 4–41. You may remember that we originally said that a spherical mirror brings parallel rays to a focus. But that statement was only approximate because of spherical aberration. The difference here is that the focus is perfect in so far as the wave nature of light can be neglected. For this reason astronomical telescopes, which have to image things essentially at infinity, often have parabolic mirrors. Similarly searchlights, which have to project a very parallel beam from a small source, often employ parabolic reflectors also.

Even the **hyperbolic mirror** (or **hyperboloid of revolution**) has special imaging properties for its foci. However, since the hyperbola is an ellipse turned inside out with one focus passed through infinity, the two foci become virtual rather than real images of each other. Furthermore, one branch of a hyperbola can be used as either a convex or concave reflector, depending on which side is silvered. As a convex mirror, the hyperbola will reflect light from the outer focus so that it appears to have come from the inner focus, as shown in Fig. 4–42. Of course the light rays in Fig. 4–42 could be turned around to show that light converging toward the inner focus but

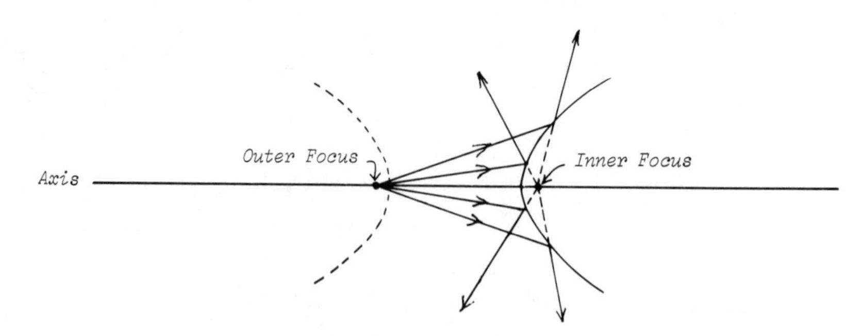

Figure 4–42 Convex hyperbolic mirror

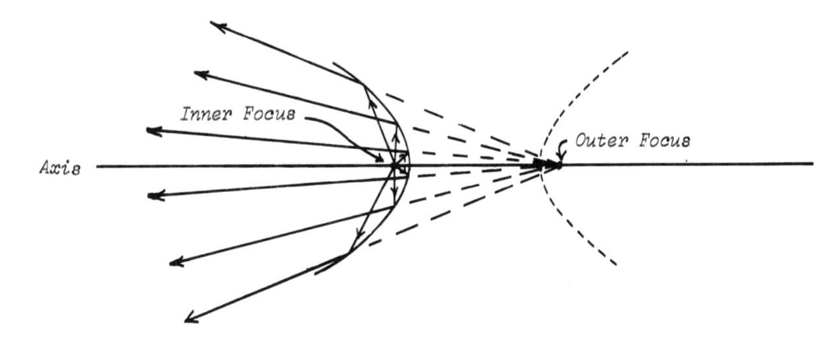

Figure 4–43 Concave hyperbolic mirror

striking the convex surface first will be reflected to pass through the outer focus. The hyperbolic mirror is often used this way as the secondary mirror in large telescopes. As a concave mirror, the hyperbola will reflect light from the inner focus so that it appears to have come from the outer focus, as shown in Fig. 4–43. Again the light rays may be reversed to show that light converging toward the outer focus but striking the concave side first will be reflected to pass through the inner focus.

Strangely enough, the first and simplest reflector which we studied, the plane mirror, can be thought of as a special case of this seemingly much more complex reflector, the hyperbolic mirror. The reason is that we can imagine a series of hyperbolas with the same foci, which approach a single straight line halfway between the foci and perpendicular to the axis. To form this set or family of hyperbolas we can go back to the ball-counterball shadows on a tabletop. Imagine the ball on the table and the point source of light level with the center of the ball. Then the counterball, the same distance away from the light source on the other side, is also touching the table, as shown in Fig. 4–44. The points of contact are the foci and the shadows give the two branches of the hyperbola. Now replace both balls with equal-sized larger ones sitting on the same points (the point light source must be raised to stay level with the centers), new hyperbolic shadows will be produced with the same foci, but they will be bigger and more open and have the two branches closer together on the axis. If this process of using successively larger balls is repeated a number of times, a whole family of hyperbolas will be produced. In the limit, the two spheres will become so large that their surfaces touch each other and the point light source, causing both shadow boundaries to merge into a single straight line perpendicular to the axis. Figure 4–45

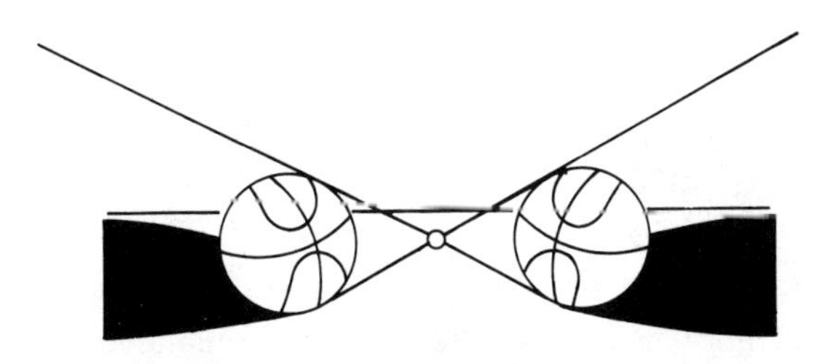

Figure 4–44 Hyperbolic shadows

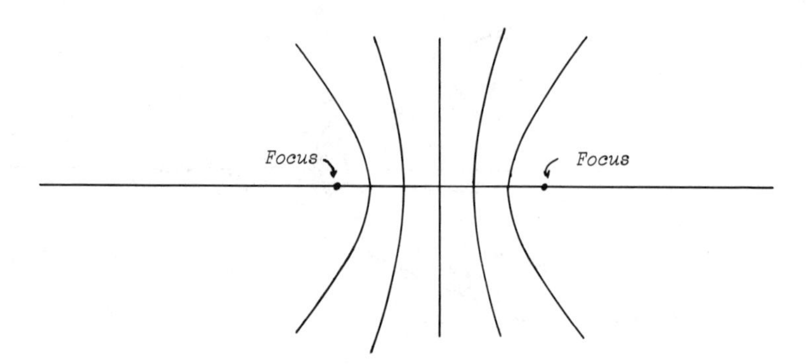

Figure 4–45 Family of hyperbolas

shows such a family of hyperbolas. Rotation of this figure around the axis gives a family of hyperboloids of revolution with a plane at the center.

Since the plane mirror can be thought of as a special case of the hyperboloid reflectors, all with the same foci, it must form a virtual image at one focus from a point object at the other focus. But the two foci are equidistant from the mirror, so it forms a virtual image as far behind it as the object is in front. Furthermore, we could have started generating the family of hyperbolas with the balls, and therefore the foci, separated by any distance and always end up with the line in the middle. So the foci for a plane mirror can be considered at any distance from it. Any point object in front of the mirror is at a focus, and therefore the mirror forms a virtual image the same distance behind at the corresponding second focus, which is precisely the same result we got much more directly by tracing two rays.

Why bother to go through all this demonstration when we already knew what the answer had to be? As a matter of fact, the plane mirror can also be considered as a special case of the spherical mirror, and from that point of view we can again prove the same result mathematically (see Appendix A). But considering the plane as a kind of hyperboloid also tells us that the imaging is perfect (focus to focus of a conic section) and without aberration. Even more generally, such a demonstration shows the complete and logical structure of geometrical optics, and what subtle, even beautiful, relationships lie buried not too far beneath the surface of things. Only a little digging and a little imagination are required of the curious mind.

QUESTIONS

1. Give an example of a situation in which a straight object appears crooked because the light rays from it do not follow a straight line.
2. Are pinhole images real or virtual?
3. How long does a full length plane mirror need to be? Draw a diagram of a person viewing rays from his or her head and feet reflected off the mirror.

4. What is the speed of light in diamond?
5. What is the index of refraction in a type of glass in which the speed of light is 1.87×10^8 m/s?
6. At a water-glass interface, which way must light be traveling in order to give total internal reflection?
7. Which has a smaller critical angle, water or ethyl alcohol?
8. Why does a diamond sparkle?
9. What does a fish in water see when it looks vertically upward? When it looks at an angle of 45.8° to the vertical? When it looks at an angle of 60° to the vertical?
10. Light rays are converging toward the back focal point of a diverging lens but strike the lens first. In what direction do they emerge from the lens? How do you know?
11. Explain the ripples of light on the bottom of a swimming pool on sunny days.
12. Draw a ray diagram for a point on an object outside the focal point of a converging lens. What is the nature of the image?
13. Draw a ray diagram for a point on an object inside the focal point of a diverging lens. What is the nature of the image?
14. Draw a ray diagram for a point on an object inside the focal point of a concave mirror. What is the nature of the image?
15. Draw an ellipse without a template and mark the foci.
16. A lamp with a cylindrical shade, placed near a wall, casts a hyperbolic patch of light on the wall above and below the shade. Can you explain why?
17. Does a spherical mirror approach the shape of a plane mirror as its radius of curvature gets smaller and smaller or as it gets larger and larger?
18. A dolphin underwater is directly under a ball held above the water surface. Does the animal see it as being closer or farther away than it really is? Draw a ray diagram.
19. Light travels from ethyl alcohol into quartz. Does it bend toward or away from the normal?
20. Would you expect more light to be reflected from a water-glass interface or a water-ethyl alcohol interface? Why?

Chapter 5

Polarization

The first reference in scientific literature to polarization effects in light was in 1670[1], about the time of Newton and Huygens. This reference concerned the strange type of refraction seen in a crystal called Iceland spar, which we now call **calcite** and which is composed of calcium carbonate. In his *Treatise on Light* of 1690, Huygens spent a good deal of effort trying to understand this strange refraction. Although he was able to explain how light must move through the crystal in order to give the unusual effect, he was totally baffled when it came to explaining why the light moved so. He thought that light was wave motion through the ether, but he considered it to be a type of longitudinal wave in which the particles of the ether pushed on each other to pass along a series of compressions. Most scientists dismissed the possibility that light could be a transverse wave in the ether, in which the particles are displaced perpendicularly to the direction of wave motion like a wave on a string, because the ether was thought to be a fluid. A fluid is not supposed to be able to support transverse waves within its body; only a solid can do that. But longitudinal waves cannot be polarized and so could not explain the effect in Iceland spar. Newton's corpuscular theory seemed to have a better chance.

The subject remained dormant for a century until a French scientist, Étienne Malus (1775–1812), again began experimenting with the crystal. In 1808 he discovered that light reflected off glass before entering the crystal behaved differently than light viewed directly through the crystal. He went on to discover a number of other facts about polarization of light and was even the first to use the word **polarization** in reference to these phenomena.

[1] Ernst Mach, *The Principles of Physical Optics: An Historical and Philosophical Treatment,* trans. John S. Anderson and A. F. A. Young (New York: Dover Publications, Inc., 1926), p. 201.

However, it remained for our two heroes of the wave theory, Young and particularly Fresnel, to work out a complete explanation in terms of waves. By 1817 Young was advancing a very tentative suggestion of transverse waves,[2] although how ether could support such waves was still unexplained. Fresnel conducted a series of interference experiments with polarized light that convinced him of the reality of transverse light waves. He even tried to describe a type of ether that would transmit these waves. After the publication of Fresnel's ideas in 1821, more and more scientists came to accept the idea of transverse light waves and it was no surprise when, some 40 years later, Maxwell's investigations showed all electromagnetic waves to be transverse waves.

To explain polarization we will refer to the three-dimensional picture of a transverse wave shown in Fig. 5-1. This wave, traveling in the x direction, is said to be **linearly polarized** or **plane polarized.** It is a wave in the electric (or magnetic, not shown) field, in which the electric field varies along a straight line in the z direction. The term *plane polarized* comes from the fact that the wave is contained in a single plane, the x-z plane. Such an electromagnetic wave would be produced by an electric charge oscillating up and down along the z direction.

Every single atom or molecule that emits light is emitting plane-polarized light instantaneously. However, any sample of light we might examine is made up from the contributions of billions of atoms, all of which are changing the polarization of their emitted waves quite rapidly. Therefore, unless we had some way of making the atoms behave cooperatively with each other, we would expect to find all possible wave orientations or polarization in equal amounts. Such light is called unpolarized and is typical of ordinary sources such as the sun, flames, and incandescent lamps. If we denote plane-polarized light by a double arrow showing the extreme values of the electric field as we view the approaching wave along the x axis, then unpolarized light could be denoted by similar arrows at all orientations perpendicular to the direction of wave motion (the x axis). This representation is shown in Fig. 5-2. In practice it becomes inconvenient to draw so many arrows, so we usually represent unpolarized

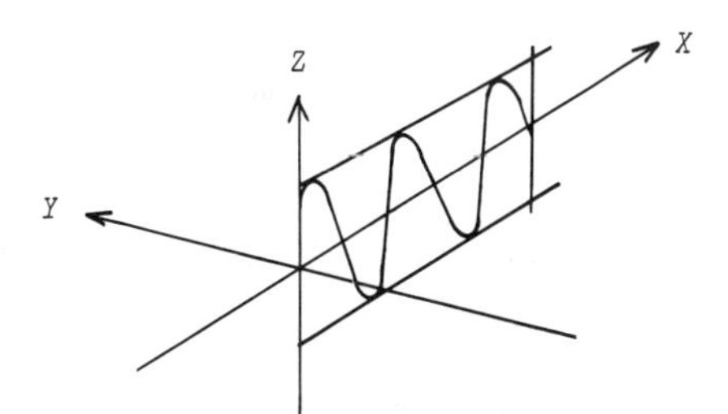

Figure 5-1 Transverse wave and polarization

[2] Ibid., p. 201; also Sir William Bragg, *The Universe of Light* (New York: Dover Publications, Inc., 1959), p. 180.

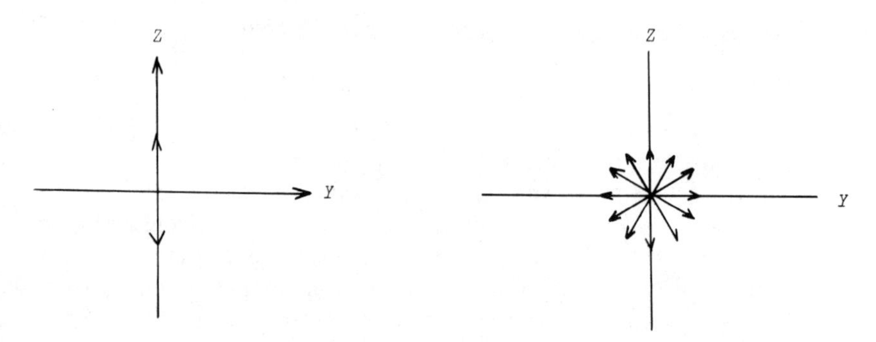

Figure 5-2 Plane-polarized and unpolarized light

light by equal-length arrows along the two perpendicular axis y and z, as shown in Fig. 5-3.

It is important to remember that such representations as Figs. 5-2 and 5-3 are merely conventional. What you would actually see as a plane-polarized wave comes toward you if you could see instantaneous electric fields, is the electric field growing from zero upward along the z axis until it reached maximum strength, then shrinking back through zero to maximum strength in the negative z direction, then growing back through zero, and so on, all with the very high frequency of light (10^{14} to 10^{15} Hz). Our eyes do not respond to this instantaneous field but rather to its average squared value, and the eye's response is indifferent to polarization. In other words, you cannot tell just from looking at light whether it is polarized or not.

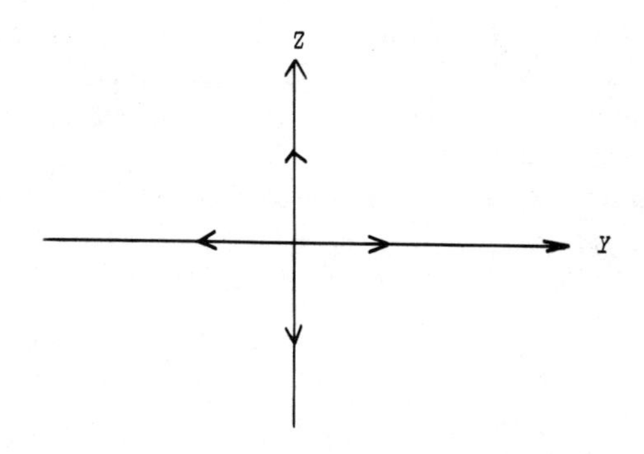

Figure 5-3 Unpolarized light

5.2 POLARIZATION BY REFLECTION

One common way of obtaining polarized from unpolarized light is by reflection. Malus discovered in 1808 that reflection of light from a nonmetallic surface, such as glass or water, produced at least partial polarization. Partially polarized light has all polarizations present but not in equal amounts; one direction of polarization is present in a greater proportion than the others. When light is reflected from a nonmetallic surface, the direction of polarization that is present in the greatest proportion in the reflected beam is that parallel to the reflecting surface. The proportion of reflected light that has this polarization depends on the angle of incidence. In particular, all the reflected light is linearly polarized in a direction parallel to the surface at one special angle of incidence, called **Brewster's angle** θ_B. The name derives from Sir David Brewster (1781–1868), a British scientist who discovered that this polarizing situation is characterized by a right angle between the reflected and refracted rays (oddly, he didn't even believe in the wave theory of light).[3] This effect is shown in Fig. 5–4.

An end view of the refracted ray of Fig. 5–4 would show that the refracted light is partially polarized, with a preferred direction perpendicular to the polarization of the reflected ray. This partial polarization must occur because both polarizations are present in equal amounts in the incident ray and only a portion of the one polarization (parallel to the surface) is reflected, leaving the remaining refracted light with a smaller proportion of that polarization.

Brewster's discovery can be explained by considering the oscillation of individual electrons in atoms as the light waves strike them. These electrons are strongly bound to atoms by electric forces and in many respects act like masses on springs, that is, they vibrate when acted upon by the oscillating force due to an electromagnetic wave. It is the vibrations of the electrons bound to the atoms of the reflecting medium that give rise to the reflected and refracted light. Since the electrons have electric charge they produce electromagnetic waves when they vibrate, which become the refracted and reflected waves. Electrons throughout the medium absorb and reradiate the refracted wave, but only those within about one-half wavelength from the surface produce the reflected wave. Now an oscillating charge produces a transverse electromagnetic wave primarily in a direction perpendicular to its own vibration; no wave is emitted along the line of vibration. This directional preference is illustrated in Fig. 5–5.

From Fig. 5–4 you can see that the electrons in the medium must be vibrating perpendicular to the reflected ray direction (parallel to the surface) and parallel to that direction if the angle between the refracted and reflected rays is 90°. The vibra-

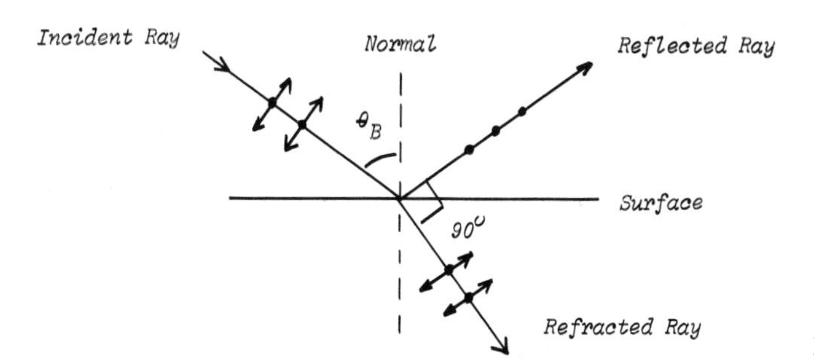

Figure 5–4 Polarization by reflection

[3] Frances A. Jenkins and Harvey E. White, *Fundamentals of Optics,* 4th ed., (New York: McGraw-Hill Book Company, 1967), p. 302.

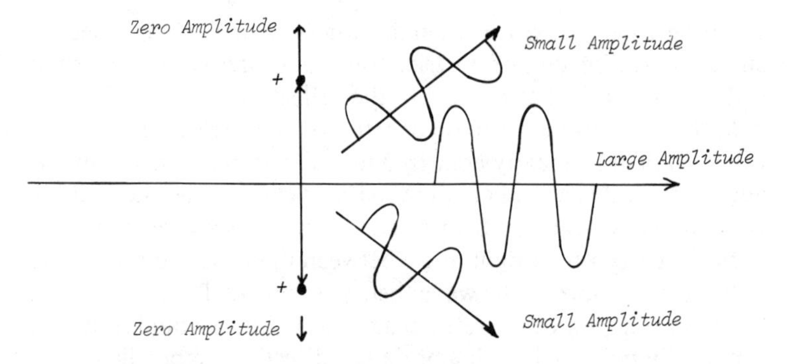

Figure 5–5 Electromagnetic wave from vibrating charge

tion parallel to the reflected ray can give no wave in that direction, as seen from Fig. 5–5, so only the electrons vibrating parallel to the surface contribute to the reflected light, and the waves must be polarized in that direction. These arguments do not hold for metals because they are characterized by free electrons, not bound to any atom, and free electrons behave differently than bound electrons when stimulated by electromagnetic waves.

It is possible to use the relation found by Brewster along with the law of refraction to obtain a formula for Brewster's angle in terms of the index of refraction of the reflecting medium by assuming the original medium has a refractive index of unity. The formula tells us that a larger refractive index gives a larger value for Brewster's angle. For water (n = 1.33) Brewster's angle is 53.1°, and for glass (n = 1.5) it is 56.3°. At these particular angles of incidence for these substances, the reflected light will be linearly polarized in a direction parallel to the reflecting surface. For other angles of incidence, the reflected light will be partially polarized with a preferred direction parallel to the surface, the polarization being more pronounced the closer the angle of incidence is to Brewster's angle.

Polarization upon reflection is the reason why polarizing sunglasses work so well to reduce glare. When unpolarized sunlight reflects off horizontal surfaces, it becomes at least partially polarized in a horizontal direction. Much of the glare, or excessive and disturbing light, is due to light reflected from such horizontal surfaces as car hoods, road surfaces, and water surfaces. The polarizing lenses of the sunglasses are oriented so that they only pass vertically polarized light and block horizontally polarized light. Therefore they block most of the light from horizontal surfaces while cutting direct light by only about 50 percent.

5.3 POLARIZATION BY SCATTERING

Another way that polarization is produced in nature is by scattering. Scattering is similar to reflection in that the light is absorbed by atoms and reradiated, but scattering means reradiation in all directions and throughout the bulk of the scattering

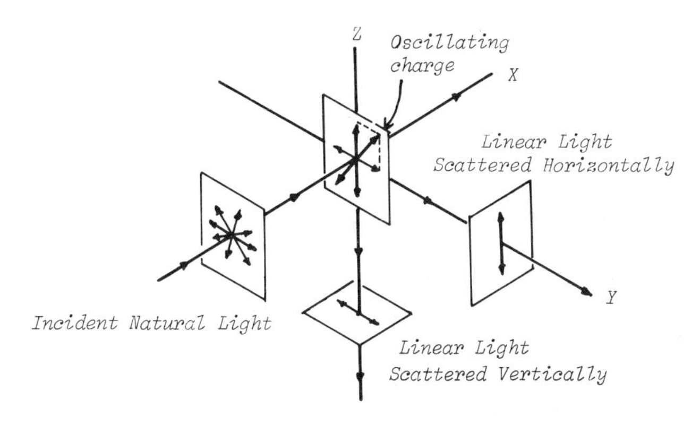

Figure 5–6 Polarization by scattering

medium, whereas reflection just takes place at a boundary. The most obvious scattering in nature is the scattering of sunlight by the atmosphere to produce the blue daytime sky. The molecules of the air scatter blue light in all directions out of the incident sunlight, so no matter in what direction we look, we see incoming blue light. The color will be investigated in a later chapter, but here we wish to note the polarization effects. Scattering at right angles tends to produce linearly polarized light. Therefore the skylight is partially polarized, with the strongest polarization arriving from a sky direction at right angles to the sun's direction. The reason for this polarization is essentially the same as the reason for polarization by reflection. The molecule that absorbs and reradiates the electromagnetic wave only can vibrate perpendicularly to the original wave direction and therefore can send only one polarization along each direction which is at right angles to the incoming sunlight, as shown in Fig. 5–6. Here the incident light is traveling along the x axis and the observer of scattered light could be along the y axis or the z axis (or any line in the y-z plane). The light along the y axis is linearly polarized in the z direction and that along the z axis polarized in the y direction. In general, the scattered light is polarized perpendicularly to the plane containing the sun-observer line and the observer's line of sight perpendicular to that.

You can observe the polarization of skylight with a polarizing filter, such as polarizing sunglasses, on a sunny day outdoors. Look at the sky along a direction perpendicular to the sun's direction. By rotating the polarizing filter in front of your eye, you should see the skylight brighten and darken as the filter turns in and out of alignment with the polarized skylight. Figure 5–7 shows the appropriate geometry.

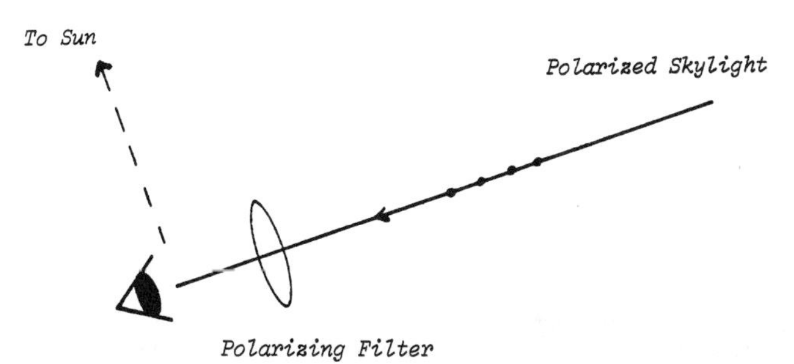

Figure 5–7 Observing skylight polarization

5.4 DOUBLE REFRACTION

We are now in a position to understand the strange polarization effects in the crystal calcite which led Huygens and Malus to investigate polarization in the first place. Calcite naturally forms a crystal that is a rhombohedron, that is, its opposite faces are parallel but there are no right angles. For simplicity of representation, in Fig. 5–8, we show the crystal with equal edges *OP, OQ,* and *OS*. Then the plane *ORTS* bisects the angle *QOP* (and *QRP*) and is called the **principal plane** or **principal section** of the crystal. What Huygens and others noted was that an incident ray of light perpendicular to the face *OPRQ* split into two rays in the principal plane, as shown. The ray passing straight through is called the **ordinary ray** because it seems to follow the usual law of refraction, while the other is called the **extraordinary ray.** If the crystal is rotated about an axis parallel to the incident ray, the extraordinary ray rotates about the ordinary ray, both remaining in the principal plane. Figure 5–9 shows this effect, looking parallel to the incident ray and the emerging rays.

Later investigators found that the two rays have different polarizations. The ordinary ray is linearly polarized perpendicularly to the principal plane and the extraordinary ray has a polarization in the principal plane, as shown in Fig. 5–10. This difference in polarization gives a clue to the physical basis for the unexpected behavior known as **double refraction** or **birefringence.** Within the crystal the atoms are arranged in a regular way and held in place by strong forces. Since any light passing through the crystal is passed from atom to atom, it is not too surprising that the crystal behaves differently in response to electric field vibrations in one direction than it does in response to those which are perpendicular to the first direction. The two perpendicular polarizations of light have, in general, different indices of refraction within the crystal: they move at different speeds. In fact, many different kinds of crystals show such double refraction, but calcite exhibits it to a larger degree.

However, different refractive indices or speeds for the two polarizations cannot be the whole explanation for the effects we see in calcite. If it were so, both the ordinary and extraordinary ray would obey the law of refraction, only with different refractive indices. In reality only the ordinary ray obeys the usual law of refraction.

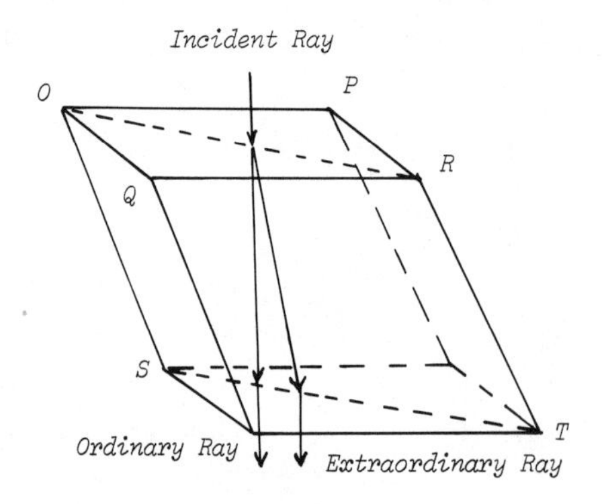

Figure 5–8 Double refraction in calcite

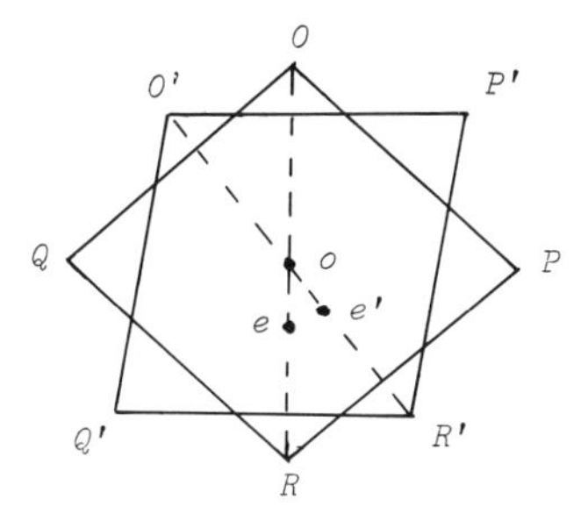

Figure 5–9 Rotation of extraordinary ray

Both polarizations strike the crystal perpendicularly in Fig. 5–8. If both obeyed the law of refraction, they would both slow down and continue straight through, although at different speeds. Instead, the extraordinary ray heads off in a different direction in the principal plane. The only way we can explain this behavior is to suppose that a wave with its polarization in the principal plane has different speeds (due to different refractive indices) in different directions within the principal plane. Thus the polarization of the extraordinary ray is characterized not by one index of refraction but by many, depending upon direction. Huygens described the situation by drawing his wavelets (see Chapter 2) for the extraordinary ray as ellipses instead of circles, with the long axis of the ellipse in the direction of fastest travel (lowest refractive index). This construction correctly predicts the direction of the extraordinary ray.

We find that there is one direction within the principal plane in which both polarizations travel at the same speed and therefore have the same refractive index. In our crystal of Fig. 5–8 or 5–10 this direction is along the diagonal OT of the principal plane, which is called the **axis** of the crystal. Of course, the ordinary ray travels with this same speed in all directions. For waves with the polarization of the extraordinary ray, the speed along the axis is the slowest speed at which they travel. These waves have a higher speed in any other direction in the principal plane and a maximum speed at right angles to the crystal axis. In this direction in the principal plane perpendicular to the crystal axis, the index of refraction of the ordinary ray n_o is 1.66 while that of the extraordinary ray n_e is 1.49. Because of this dependence on direction

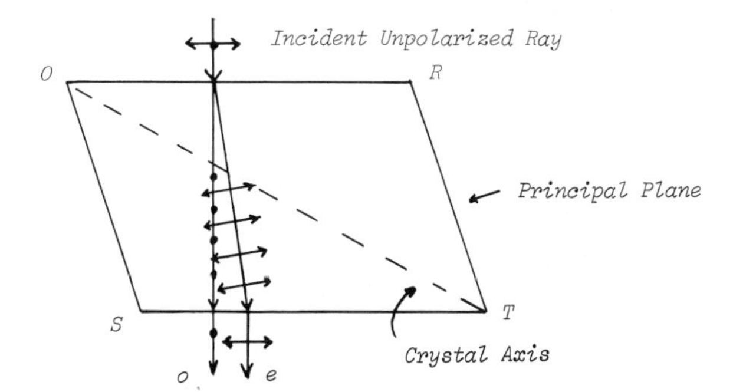

Figure 5–10 Polarization of rays

birefringent materials are also called **anisotropic.** Isotropic means the same in all directions; anisotropic means not isotropic or not the same in all directions. In summary, the effect of a birefringent crystal upon an unpolarized incident light ray is to break it into two rays, the ordinary ray linearly polarized perpendicular to the crystal's principal plane and the extraordinary ray linearly polarized in that plane.

5.4.1 Dichroic Crystals

Certain crystals, such as tourmaline, not only show double refraction but also absorb one of the polarizations much more strongly. They are said to be **dichroic.** This effect is illustrated in Fig. 5–11. As you can see, such a crystal provides a convenient method for obtaining plane polarized light from unpolarized: simply pass it through the crystal. Unfortunately, it is quite difficult to grow and maintain large, optically fine crystals in order to produce a large polarized beam. This difficulty was first overcome in 1934 by the young American scientist Edwin Land (1909–), who was trying to find a way to produce large, flat sheets of polarizing material. He found that certain small dichroic crystals could all be aligned the same way in a flow process and that when the liquid was evaporated, a sheet of polarizing material was left. Later he discovered that a certain plastic became doubly refracting when stretched and became dichroic when stained with iodine. These sheets of polarizing material he named Polaroid and this was the beginning of the Polaroid Corporation, later famous for self-developing film and camera. We shall have occasion to refer to Edwin Land again when we consider how we see color. Nowadays when we wish to obtain polarized light we usually just pass unpolarized light through a polarizer such as a Polaroid sheet; the desired plane of polarization is obtained by rotating the axis of the polarizer. Such an arrangement is shown in Fig. 5–12.

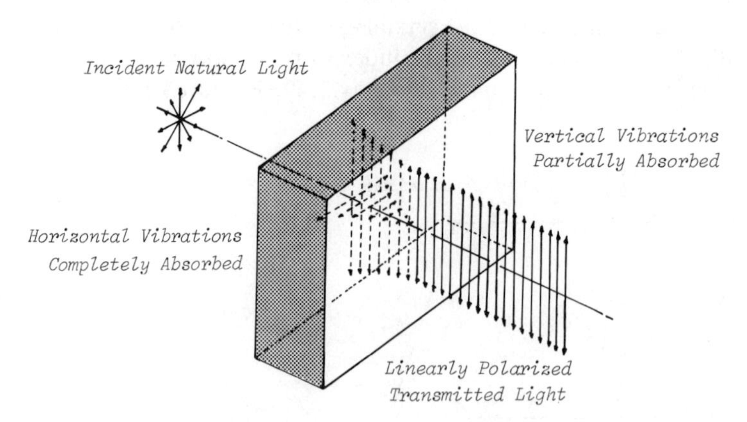

Figure 5–11 Dichroic crystal

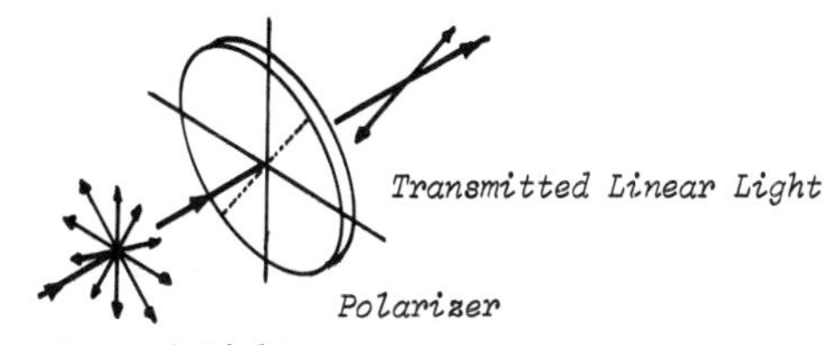

Figure 5-12 Polarizing filter

5.4.2 Retardation Plates

Some remarkable and beautiful effects can be produced by combining polarizing filters and birefringent crystals. First we must produce what is called a **retardation plate** by cutting out a section of a birefringent crystal, such as calcite, that has two faces parallel to the crystal axis, as shown in Fig. 5-13. Any light passing through the retardation plate perpendicularly to its large faces will also be moving perpendicularly to the crystal axis, in which direction there is a maximum difference in the speeds of the ordinary and extraordinary rays. Furthermore, both polarizations will follow the same path straight through if normally incident on one face. That is, the ordinary and extraordinary rays will not separate when incident perpendicular (or parallel) to the crystal axis. Because of the difference in speeds the polarization characteristic of the ordinary ray (in the case of calcite) will fall behind that of the extraordinary ray and is said to be retarded—hence the name retardation plate. How much one polarization is retarded with respect to the other upon emergence depends on the thickness of the plate.

By combining the proper thickness retardation plate with a polarizing filter we can produce a so-called half-wave plate, which turns the original direction of polarization through 90°. Figure 5-14 illustrates this effect. Here the original unpolarized light, (*a*), is first incident on a polarizer, (*b*), with its axis at a 45° angle to the axis of the retardation plate (calcite crystal axis). The effect of the polarizer is to ensure that the ordinary and extraordinary polarizations in the crystal start out in phase and with equal amplitude. As far as the retardation plate is concerned, the linear polarization at 45° to its axis is equivalent to two equal amplitude waves, one

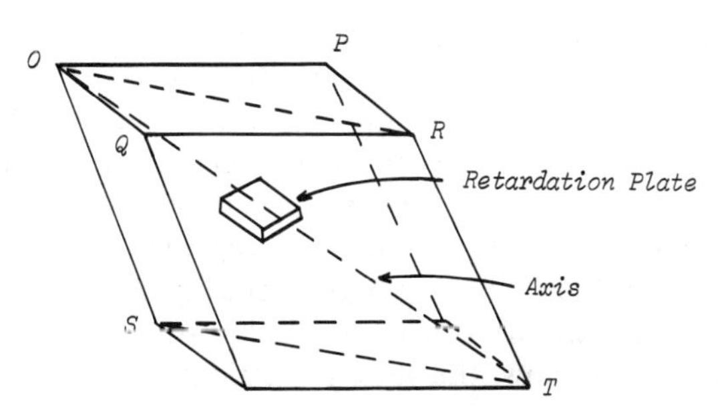

Figure 5-13 Cutting a retardation plate

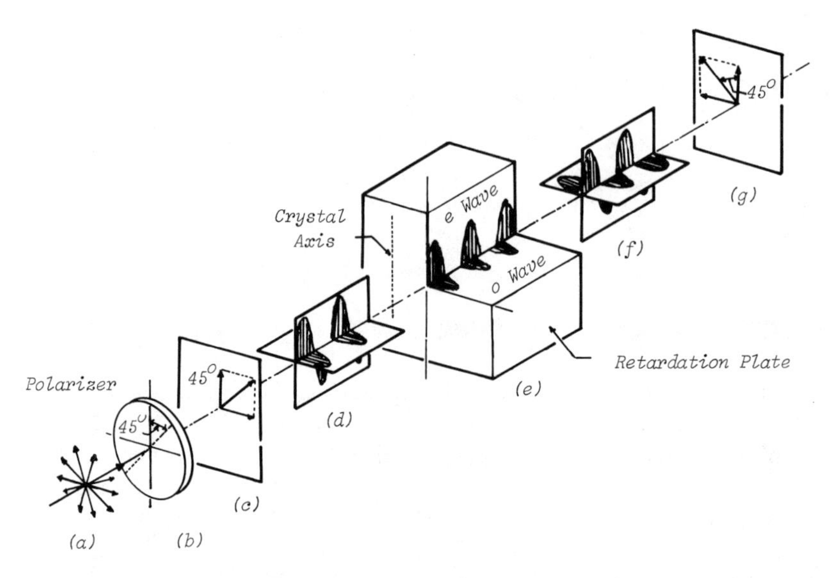

Figure 5–14 Action of a half-wave plate

(extraordinary) parallel to its axis and one (ordinary) perpendicular to it, as shown at steps (*c*) and (*d*) of Fig. 5–14. Note in (*d*) that both waves have zeros at the same place and that the extraordinary wave has a maximum upward where the ordinary wave has a maximum to the *right.* However, in the crystal this relation between the two waves changes because the extraordinary polarization moves through faster. In other words, the extraordinary polarization has a longer, and the ordinary polarization a shorter, wavelength in the crystal. The crystal is just thick enough so that the ordinary polarization is retarded by a half wavelength with respect to the extraordinary polarization, as shown in steps (*e*) and (*f*) of Fig. 5–14. Now you can see in step (*f*) that where the extraordinary wave has a maximum upward, the ordinary wave has a maximum to the *left.* Step (*g*) of Fig. 5–14 shows that this result is equivalent to a linear polarization 45° from the crystal axis and 90° from the original polarizer axis. The retardation plate has turned the plane of polarization through 90°. Of course, a retardation plate of a single thickness can be a perfect half-wave plate for only one wavelength (one spectral color).

For convenience quite often we represent the situation of Fig. 5–14 with just the schematic input and output planes (*c*) and (*g*), as in Fig. 5–15. In this figure, the two arrows at right angles represent the simultaneous maximum values of the electric fields in the ordinary and extraordinary waves. We must keep in mind that if we were seeing the electric fields over a period of time as the waves moved toward us, the arrows of Fig. 5–15 would oscillate in length through zero together. We can get from the input to the output picture in Fig. 5–15 simply by replacing the arrow of the ordinary wave with the arrow representing the electric field in the ordinary wave one half wavelength ahead. Figure 5–16 shows the output electric fields in time sequence, as discussed above.

Figure 5–15 Input and output of half-wave plate

This half-wave plate effect can give rise to striking and beautiful polarization colors. For example, a jumble of small, clear crystals grown on a glass plate will often show a remarkable array of colors when placed between crossed polarizers and viewed by transmitted white light. The crossed polarizers together without the crystals between would transmit no light. But the crystals of random thickness and orientation can act as half-wave plates at some positions for some wavelengths. Then at those positions the appropriate wavelengths will have their original plane of polarization turned through 90° to alignment with the last polarizer and so be allowed through. Thus one sees a rather random array of colors. Actually the two polarizers could even be aligned; then any wavelength with its plane of polarization turned through 90° will be extinguished by the final polarizer, giving rise to the complementary color at that position. However, the effect is usually more striking with crossed polarizers because that configuration produces a dark background against which the polarization colors are seen.

5.4.3 Induced Birefringence

Even isotropic materials can often be made anisotropic, or birefringent, and so can also show polarization colors. Materials such as glass and plastic have no underlying crystal structure and therefore would normally be isotropic. However if these materials are stressed by the application of large forces to them at various points, they become doubly refracting. This effect can be seen by viewing the stressed

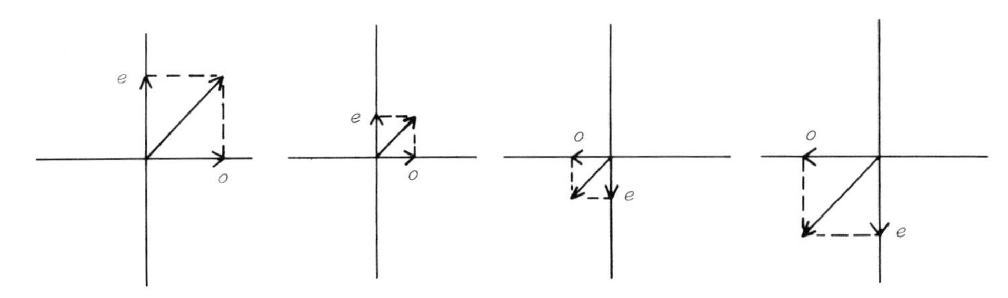

Figure 5–16 Time sequence of output

material placed between crossed (or aligned) polarizers with transmitted light; areas of transmitted colored light will clearly show the areas of stress in the material. Often components of machinery or even architectural structures are tested by making models out of clear plastic and then viewing the model, loaded as in actual use, with polarizing filters. The areas of greatest stress can be clearly seen in this manner. You can even see similar results yourself with two polarizing filters (such as the lenses from the polarizing sunglasses) and a cheap clear plastic box or similar item. Usually the process of forming a cheap plastic item builds in permanent stresses. When the item is placed between crossed polarizers and viewed with transmitted light, beautiful polarization colors appear streaked throughout it. If one polarizer is rotated, the colors change. You may have even seen a similar result wearing polarizing sunglasses and looking through a car window. The light originally incident on the window could be partially polarized by skylight scattering or by reflection. The car window itself is usually under stress, particularly if it is curved, as a result of its installation. Then through the final polarizer of your sunglasses you can see streaks of color in the window. Some materials can even be made doubly refracting by application of large electric or magnetic fields. The names given to these phenomena are the **Faraday effect,** the **Kerr effect,** or **Pockel's effect,** but they are really beyond the scope of this text.

5.5 CIRCULAR POLARIZATION

A quarter-wave plate can be made in a manner similar to that used for a half-wave plate, but with the retardation plate half as thick. This device gives rise to a type of polarization we have not yet considered. In this case the retardation of the ordinary wave is a quarter wave. The resulting input and output time sequence are shown in Fig. 5–17.

A quarter wave in front of the ordinary maximum to the left (wave approaching) is a zero, shown in the first output frame of Fig. 5–17 with the upward maximum of the extraordinary wave. The resulting electric field is just upward. In

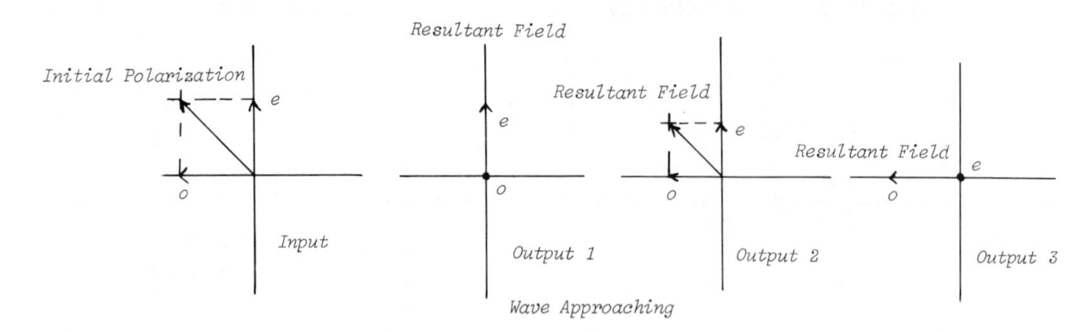

Figure 5–17 Action of a quarter-wave plate

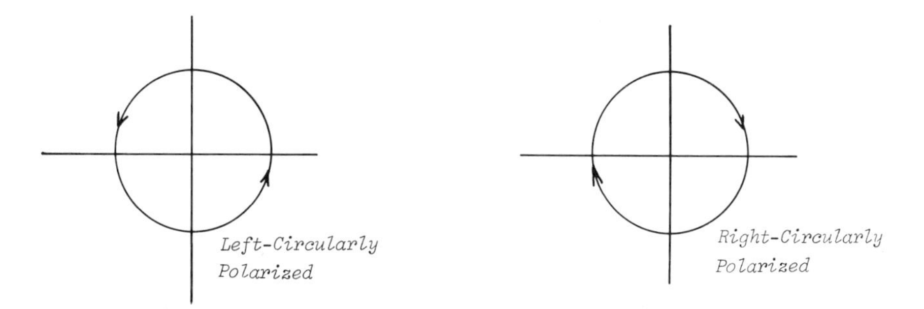

Figure 5–18 Representation of circularly polarized light

the second output frame the ordinary electric field has begun to grow toward its left-hand maximum while the extraordinary electric field has shrunk toward zero, and the resultant electric field is tilted to the left. The third output frame shows the ordinary maximum to the left, which was originally aligned with the upward extraordinary maximum of the first frame. Now it has slipped back one quarter wave to alignment with the following extraordinary zero, giving a resultant electric field to the left. Something quite different is happening here: the resultant electric field is not oscillating along a line as the waves approach but rather is keeping the same length while spinning in a circle. Light which behaves in this way is said to be **circularly polarized.** When the electric vector of the approaching wave spins as in Fig. 5–17, it is called **left circularly polarized.** Spinning in the opposite sense would give **right circularly polarized** light. Right circularly polarized light would result if the extraordinary wave were retarded one-quarter of a wavelength with respect to the ordinary wave or if the initial polarizer were turned through 90°.

To show circularly polarized light in one picture we usually show the path of the tip of the electric field arrow, as in Fig. 5–18. We can make a perspective drawing of a right circularly polarized light wave omitting the two linearly polarized components, which can be thought of as adding up to the resulting wave (Fig. 5–19). This might clarify what is physically happening. You can see that the three-dimensional picture of a circularly polarized wave looks like a spiral.

A quarter-wave plate (initial polarizer at 45° followed by a retardation plate) changes unpolarized incident light into circularly polarized light. This circularly polarized light does not look any different from unpolarized light to the naked eye. Even if it is viewed with another polarizing filter, it looks the same as unpolarized

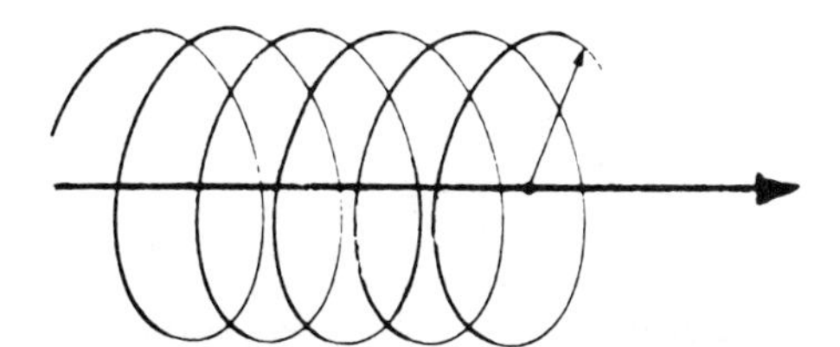

Figure 5–19 Right circularly polarized wave

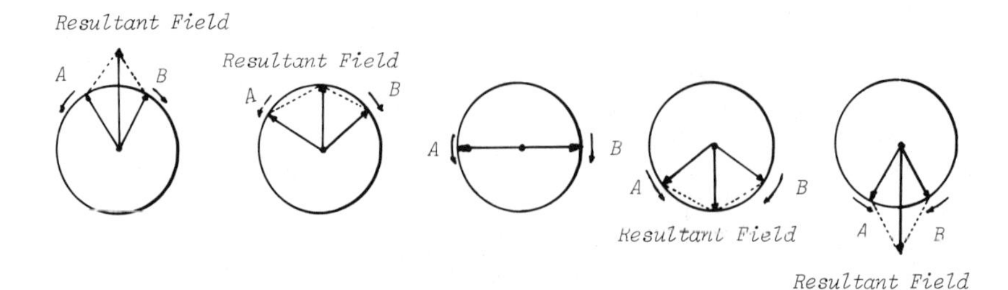

Figure 5–20 Linear polarization from two circular polarizations

light, there being no change in intensity as the polarizer is rotated. However, circularly polarized light can be detected by viewing through another quarter-wave plate. If the retardation plate is turned toward the incoming light to be tested and the polarizer toward the eye, the retardation plate will cause another quarter-wave retardation of one polarization, and as a result linearly polarized light will strike the final polarizer. Now as the device is turned, the intensity will increase and decrease according to whether the linearly polarized light and final polarizer are more or less aligned. Unpolarized light would show no such variation viewed through a quarter-wave plate in this fashion.

We introduced circularly polarized light by a kind of adding together or superposition of two mutually perpendicular linear polarizations (see Fig. 5–17). However, it is just as valid to consider linear polarization as a superposition of a right and a left circularly polarized light. Figure 5–20 shows how this can be done in a time sequence.

To obtain a different linear polarization, we simply need to adjust the relative phase of the two circular ones. That is, we start in the first frame with electric field A in a different position and then let both A and B rotate at the same speed. This viewpoint is particularly applicable to the photon concept of light. Photons should be thought of not only as moving at the speed of light but also as spinning about an internal axis, much as the earth or a top spins. The spin axis of a photon can only be pointed along its direction of motion or opposite to that direction. A beam of circularly polarized light is the same as a stream of photons all spinning the same way, as in Fig. 5–21. Then, in this interpretation, linear polarization would be a superposition of both circular polarizations. We must remember, however, that this superposition is a rather mysterious quantum mechanical one, in which *each* photon must be thought of as having both spin states at the same time. The point is that in terms of photons circular polarization is the more basic state and linear polarization is the more complex one.

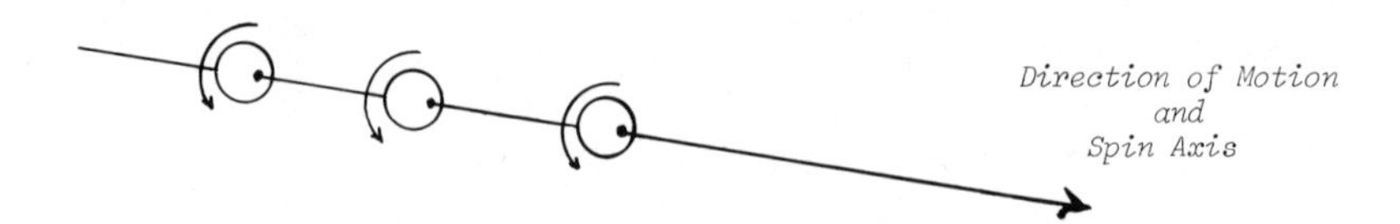

Figure 5–21 Photon interpretation of circular polarization

QUESTIONS

1. Describe the difference between a longitudinal and a transverse wave. Which can be polarized?
2. Who discovered polarization by reflection?
3. Who was the first to suggest that light was a transverse wave?
4. In terms of waves, what is the difference between plane-polarized and unpolarized light?
5. What is Brewster's angle?
6. If unpolarized light is incident on the surface of a body of water at Brewster's angle (53.1°), what is the angle of reflection? What is the angle of refraction? (A diagram will help you.)
7. What are the states of polarization of the reflected and refracted rays in question 6?
8. Explain how polarizing sunglasses preferentially reduce glare.
9. In what direction should you look in the sky to most easily detect the polarization of skylight?
10. Why does light become polarized by reflection?
11. Why does light become polarized by scattering?
12. You have set a calcite crystal down over a period on a printed page and you can see two dots through the crystal. How can you discover which image corresponds to the ordinary ray and which to the extraordinary?
13. In a calcite crystal what is the speed of the ordinary wave in a direction perpendicular to the crystal axis in the principal plane? What is the speed of the extraordinary wave in the same direction? What are the speeds of the two waves along the crystal axis?
14. Given a clear plastic box and two polarizing filters, how would you demonstrate polarization colors?
15. What causes polarization colors?
16. What is circularly polarized light? How can it be produced and detected?
17. In terms of photons, what is circularly polarized light? Linearly polarized light? Unpolarized light?

Chapter 6

Lasers

A revolutionary advance in the generation of light occurred in 1960 with the invention of the **laser.** The word *laser* is an acronym for *l*ight *a*mplification by *s*timulated *e*mission of *r*adiation, which is a fair description of how a laser works. A similar device called a **maser,** which is the same acronym but with *microwave* substituted for *light,* had been invented in 1954 by Charles Townes (1915–) at Columbia University. Townes shared the 1964 Nobel prize in physics with two Soviet scientists, who had independently thought of the same principle. As you might imagine, a maser has an output in the microwave region of the spectrum. Almost as soon as the maser was invented, scientists began trying to extend the principles involved to the much shorter wavelengths of light. In 1958 Townes and Arthur Schawlow (1921–), who was then at Bell Telephone Laboratories, published a paper establishing the basis for "optical masers," as lasers were originally called. Then in 1960 Theodore Maiman (1927–) at the research laboratories of Hughes Aircraft Company built the first operational laser from a ruby crystal. From that time to the present there has been an amazing proliferation of types, capabilities, and applications of lasers. More than any other single discovery, this device has led to a rebirth of the whole science of optics.

6.1 ABSORPTION AND EMISSION

To understand how a laser works and what its special properties are, we need to return to the basic interactions of light with matter at an atomic level which we studied in Chapter 3. There we noted that individual atoms had discrete energy levels

Figure 6–1 Absorption and spontaneous emission

in which they could be and between which they could jump, with the emission or absorption of the appropriate amounts of energy. When atoms combine in a molecule or are held in a crystalline solid, similar rules apply except that there are then often energy bands in which the atoms can exist. An energy band can be thought of as a great number of very closely spaced energy levels, so that in effect the atom or molecule can have any energy within the band. We can illustrate such results diagrammatically by drawing horizontal lines and bars for energy levels and bands, with higher energies higher in the diagram. Then we can show the state of any atom by placing a dot in the appropriate level or band. Thus, simple absorption and emission of a light wave would be shown as in Fig. 6–1.

We know from Chapter 3 that the energies of the atom and the light frequency must be related for these transitions to occur:

$$hf = E_1 - E_0 \tag{11}$$

The emission process shown in Fig. 6–1 is called **spontaneous emission** because it will occur without any outside influence on the atom shortly after the atom reaches the higher energy level. The normal state for the atom to be in is in the lowest possible energy, which is called the **ground state.** Typically, spontaneous emission will occur within one-millionth (10^{-6}) of a second after the atom has been excited to a higher energy level; in some cases the average stay in the excited (higher) level is as short as 10^{-8} s. Ordinary light sources such as electric lamps, flames, or the sun emit light by spontaneous emission.

However, it had been known for a long time before 1960, or even 1954, that another type of transition is possible between the energy levels of Fig. 6–1. In 1917 Einstein showed that if an atom is in a higher energy level, E, it can be stimulated to emit by an incoming light wave of just the right frequency, as given by Eq. (11). Figure 6–2 illustrates. This process of stimulated emission is notable because the emitted light has exactly the same frequency, direction, and polarization as the stimulating light. The emitted light is even in phase with the stimulating light, lining

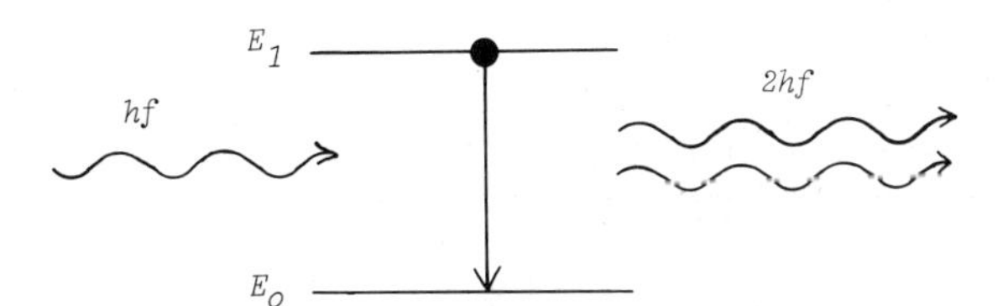

Figure 6–2 Stimulated emission

up crest to crest and trough to trough. In other words, the emitted light simply adds to the power of the incoming light without changing it in any other way.

6.2 POPULATION INVERSION

Stimulated emission is the basic process underlying the operation of all lasers. But more is needed to make a laser work. For one thing, many atoms must act together in a laser to produce a beam with any appreciable power. To get many atoms (or molecules) to undergo stimulated emission, we must start with many in an **excited state** above the ground state. In fact we must start with more in the upper state than in the lower state, a condition known as an **inverted population.** The reason for this prerequisite is that light of the correct frequency is just as likely to excite a lower state atom that it strikes as to stimulate emission from an upper state atom that it strikes. If there were not more upper than lower state atoms in the medium when a light wave started through, then it could gain no net energy; any energy it picked up by stimulated emission would be more than compensated for by energy lost to absorption by the lower state atoms.

How then can we obtain an inverted population in some medium in order to start laser action? Light input of the correct frequency can excite some of the atoms of a medium but we could never reach an inverted population by using a preliminary flash of light, just working between two levels as in Fig. 6–1 or 6–2; as soon as 50 percent of the atoms were excited, the remainder of the preliminary flash would stimulate as many of those to emit as it excited others, so the ratio would stay at 50 percent. We can, however, obtain an inverted population with a preliminary flash of light by working with three levels instead of two. This is precisely what Maiman did with the first ruby laser. Using light to excite enough atoms for an inverted population is called **optical pumping.**

6.3 RUBY LASER

Ruby is a crystal of aluminum oxide (Al_2O_3) doped with chromium (Cr^{3+}) ions. That means that every so often in the crystal, the position at which an aluminum ion would normally be is occupied by a chromium ion (these ions are atoms missing three electrons). It is the chromium impurities that make the crystal optically interesting: without the impurities the crystal would be colorless. The more chromium, the deeper red the resulting ruby. In this respect it is interesting to note that if the aluminum oxide crystal is doped with a different impurity such as titanium or iron, sapphires of various colors result. Figure 6–3 shows the energy levels and transitions involved in

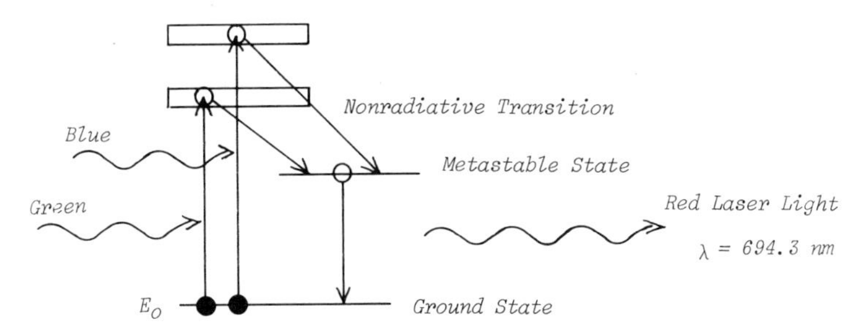

Figure 6–3 Energy levels of Cr^{3+} in ruby

laser action for chromium ions in ruby. The incoming light raises the atom to one of two energy bands. Actually, white light is used for optical pumping but the Cr^{3+} ion will only absorb the appropriate green or blue wavelengths corresponding to the two jumps shown. It is just this absorption of green and blue light that gives ruby its red color in the first place. The fact that the Cr^{3+} ion has two rather broad bands to jump to improves the efficiency of the optical pumping; more of the white pump light is effective in producing transitions. From these bands the ions drop back to an intermediate energy level which is **metastable.** This state is called *metastable* because the ions stay in it an unusually long time; on the order of milliseconds (10^{-3} s), which may not seem long to us but is a long period on the atomic scale. The drop from a band down to a metastable state is called a **nonradiative transition** because no electromagnetic radiation is emitted in the process. Instead the Cr^{3+} ion gives up some energy to atomic vibrations of the crystal to make the jump; in other words, the crystal gains heat energy. If the pumping light is very intense, then an inverted population can be formed with more ions in the metastable state than in the ground state. The reason that this scheme works is because the ions rapidly drop from the bands down to the metastable state but stay there a relatively long time. Thus, they pile up in the metastable state, a population inversion occurs, and laser action is ready to start.

One more element is needed to complete our laser: an optical cavity formed by two small mirrors, one on either side of the cylindrical ruby crystal, as shown in Fig. 6–4. The optical cavity serves to collimate the laser light into a very parallel beam and

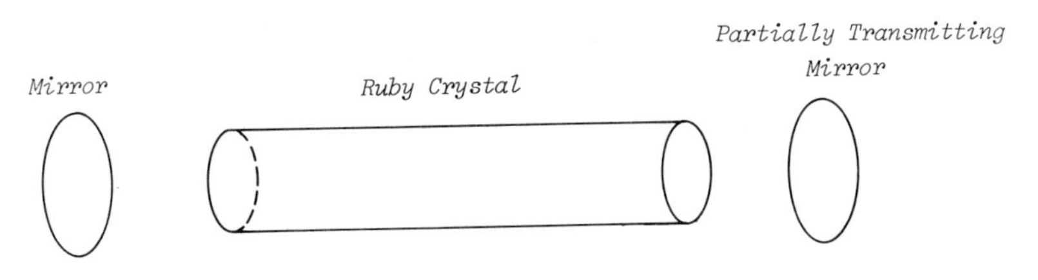

Figure 6–4 Laser optical cavity

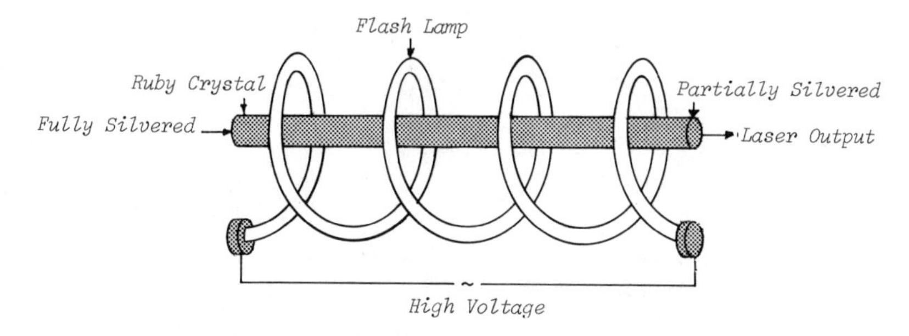

Figure 6–5 Ruby laser

to narrow the output band of wavelengths or increase the purity of the laser light. We will return to these effects in a moment, but first let us trace the buildup of the laser pulse. During the preliminary flash of white light, an inverted population is rapidly built up. At some point while the population inversion is present, one of the Cr^{3+} ions in the metastable state will spontaneously emit red light (λ = 694.3 nm) along the axis of the crystal, thereby dropping back to the ground state. This light wave travels through the crystal, stimulating other ions in the metastable state to emit waves perfectly in step. The stimulating wave gains in power in this manner, while the end mirrors reflect it back and forth through the active medium. At each incidence on the front mirror, a portion of the wave is transmitted to become the laser beam we see. The laser light continues to built until the population inversion is depleted and then it cuts off. What we end up with is a pulse, with about 0.5 ms duration, of intense, parallel monochromatic light.

Quite often the end mirrors are placed right on the ends of the ruby crystal, as in Fig. 6–5, which shows a typical ruby laser with the flash tube for optical pumping wrapped around the ruby. Another method of efficiently getting the pump light into the ruby is to use a straight flash tube at one focus of an elliptical reflecting cylinder, with the ruby rod at the other focus. This arrangement is shown in Fig. 6–6.

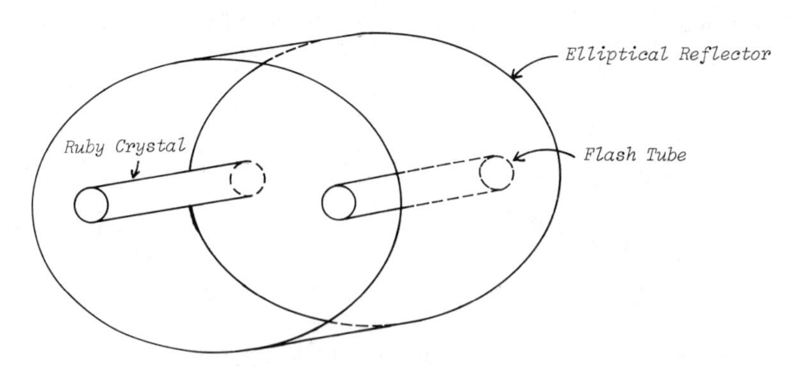

Figure 6–6 Elliptical reflector arrangement

6.4 HELIUM–NEON LASER

Of course a ruby laser is not the only type of laser. At room temperature a ruby laser will only emit short bursts of laser light, each laser pulse occurring after a flash of the pumping light. It would be very nice to have a laser that emits light continuously. Such a laser is called a **continuous wave** (CW) laser. The first CW laser was built in 1961 by Ali Javan at Bell Telephone Laboratories. This laser uses a mixture of the gases helium and neon as its active medium and is presently the most common type. Population inversion in the helium-neon laser is not produced by optical pumping at all. Instead, energy is introduced into the active medium by an electric discharge down the length of a glass tube containing the gases. The electrons in the electric discharge collide with and excite the helium atoms, which in turn collide with and excite the neon atoms. It is a transition between energy levels of the neon atoms that gives rise to the laser light. Figure 6–7 shows the transitions involved.

The first transition shown in Fig. 6–7 is the excitation of helium by the electric discharge. It so happens that neon has an energy band at just the same level as the upper helium level. Therefore, when a ground-state neon atom collides with an excited helium atom, it is easy for the helium to give up its energy to the neon, as shown. Furthermore, the helium upper level is a metastable state, which means that the atom will stay in that state for a relatively long time without spontaneous emission and thus have ample opportunity for an energy-transferring collision with a neon atom. The excited neon atoms first fall to a lower excited state by emitting red light (λ = 632.8 nm)—this is the laser transition. The neon atoms continue on to the ground state through radiative and nonradiative transitions. It is important for the cw operation of a helium-neon laser that the laser transition *not* be the one to the ground state. When the electric discharge starts, there are almost no neon atoms in the lower state involved in the laser transition, and therefore it is very easy to reach an inverted population between the two levels of interest. As laser action proceeds, the lower level involved is rapidly emptied by further transitions so that it is possible to sustain an inverted population and laser light output with a modest electrical power input

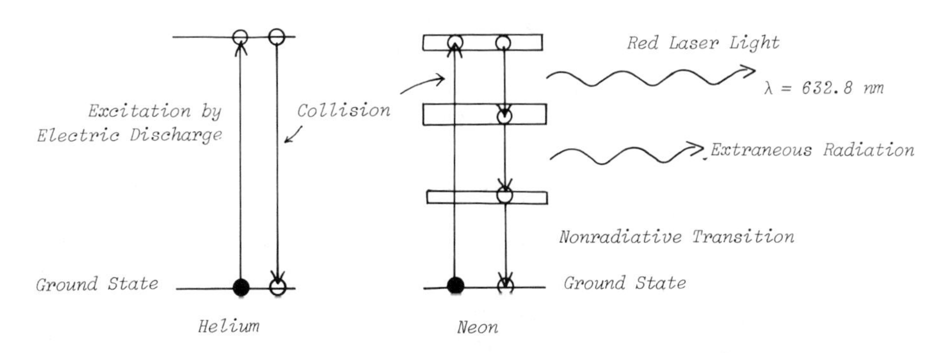

Figure 6–7 Transitions in helium-neon laser

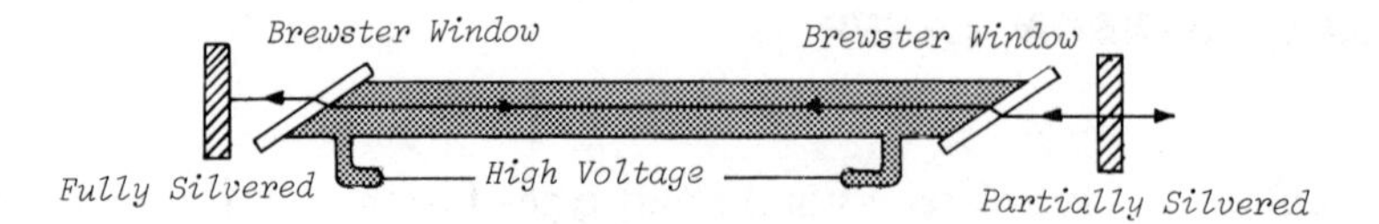

Figure 6–8 Helium-neon laser

(the continuing electric discharge). Addition of the end mirrors of the optical cavity gives the helium-neon laser as shown in Fig. 6–8.

The Brewster's angle windows on the ends of the tube in Fig. 6–8 serve to keep the laser light in one constant linear polarization. Any light reflected at this window must have a polarization parallel to the window surface and perpendicular to the plane of the drawing. Therefore, the polarization in the plane of the drawing is passed 100 percent and is the polarization of the wave that can build up by repeated passes through the medium. Correspondingly, this latter polarization becomes the polarization of the output beam. However, not all helium-neon lasers have Brewster's angle windows, and those that do not will emit unpolarized light.

6.5 OTHER LASER TYPES

Ruby and neon are just the first two of many substances that have been used as the active medium of a laser (said to have been made to "lase"). There are literally hundreds of materials that have been made to lase at thousands of wavelengths throughout the ultraviolet, visible, and infrared spectra. For example, any substance that will fluoresce is a likely candidate for the active medium of a laser. Fluorescence is the process in which a material absorbs radiation at one frequency and reemits the energy at a lower frequency (longer wavelength). This process gives a particular brilliance to the emitted color because there can be more power at that wavelength in the light returned from the substance than there was in the incident radiation. Many organic substances exhibit fluorescence, and the brilliant colors of some flowers, such as azalea blossoms, are due to this effect. Ruby itself is fluorescent because, as we have seen, it absorbs blue and green light and emits red. Some materials, notably the inert gases argon and krypton, can be made to lase at several widely separated wavelengths, any one of which can be selected with a prism. Lasers made with the gas carbon dioxide as the active medium can have very high efficiencies: they may convert up to 30 percent of the input energy into laser radiation output, compared with 0.02 percent or less for an ordinary helium-neon laser. For this reason some extremely powerful carbon dioxide lasers have been made. The infrared output from such a laser has a power that depends on the length of the tube or the total amount of active material. A 1-m long carbon dioxide laser may produce a CW output power of 100 W CW, whereas the comparable size helium-neon laser would only produce 1 to 10 mW (0.001 to 0.01 W). Huge carbon dioxide lasers have produced a CW output of

TABLE 6-1 CHARACTERISTICS OF SOME COMMON LASERS

Active medium	Wavelength (nm)	Type
Helium-cadmium	441.6	CW
Argon	476.5, 488.0, 514.5	CW
Krypton	476.2, 520.8, 568.2, 647.1	CW
Helium-neon	632.8	CW
Ruby	694.3	Pulsed
Gallium arsenide	840–904* (IR)	CW
Neodymium	1060 (IR)	Pulsed
Carbon dioxide	10,600 (IR)	CW

* Depends on temperature.

thousands of watts. Pulsed lasers can produce an output of thousands or even millions of watts by reducing the duration of the output pulse to correspondingly small times (microseconds and shorter), but their average power outputs over time intervals of several seconds are not impressive. Table 6–1 shows some of the more common lasers and their characteristics.

6.6 EFFECT OF THE OPTICAL CAVITY

So far you have seen that the mirrors of the optical cavity help the laser operation by repeatedly sending the light back through the active medium to stimulate more emission. But the contribution of the mirrors is more profound than that. For one thing, they ensure that the laser light will be collimated, because any emitted at even a small angle of inclination will walk off the mirrors after a few passes, as shown in Fig. 6–9. Any such light cannot build up by stimulated emission significantly since it does not pass through the active medium enough times. Instead, light parallel to the axis soaks up all the power because it continues to transverse the active medium time after time. This light parallel to the axis is what we observe in the output beam. One can see from Fig. 6–9 that to reduce the number of times any nonaxial light passes through the active medium and thereby reduce the chance for its buildup through stimulated emission, the size of the end mirrors should be small compared with their separation.

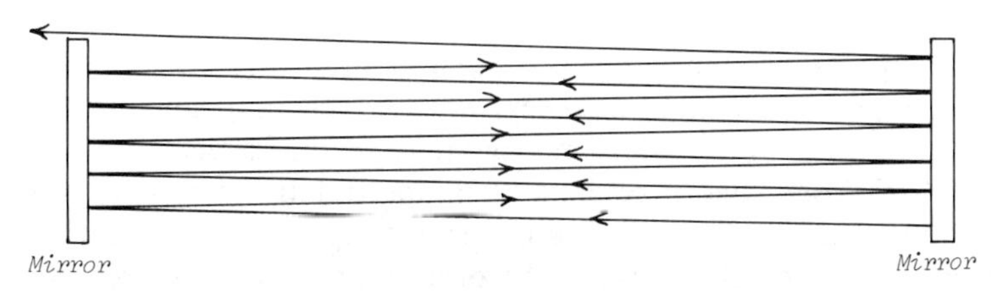

Figure 6–9 Mirror walk-off

6.6.1 Linewidth

The optical cavity also serves to decrease the range of frequencies (or wavelengths) emitted by the laser. Up to this time we have spoken of the light emitted when an atom or molecule falls from higher to lower energy levels as being of a single wavelength or frequency. Now we must face the fact that that is not quite true: there is always a small range or band of frequencies, often spoken of as the **bandwidth** or **linewidth.** If we assume that there are no outside disturbances, the spontaneous emission of an atom is characterized by a natural linewidth, Δf_n, which is related to the **natural radiative lifetime** Δt_n, of the transition. The natural radiative lifetime is the average time the atoms spend in the higher energy state before dropping to a lower state and emitting radiation. It is impossible to predict precisely when an individual excited atom will emit, but there does exist an average lifetime for a collection of identical excited atoms. This natural radiative lifetime is therefore roughly the time interval over which the collection of excited atoms will be emitting radiation spontaneously if they are assumed to all act independently and remain free of outside influences. The approximate relationship that always holds between linewidth and lifetime is

$$\Delta f_n \simeq \frac{1}{\Delta t_n} \tag{12}$$

Example:

If the natural radiative lifetime of a transition between energy states of an atom is 2×10^{-8} s, what is the natural linewidth of the spontaneously emitted radiation?
We are given $\Delta t_n = 2 \times 10^{-8}$ s.
Substitution into Eq. (12) gives

$$\Delta f_n = \frac{1}{(2 \times 10^{-8})\text{s}} = 0.5 \times 10^{-8}\,\text{s}^{-1}$$

$$\Delta f_n = 5 \times 10^{7}\ \text{Hz}$$

It is important to remember that the linewidth Δf refers to the *range* of frequencies emitted, not the actual frequency, which is generally much larger.

The natural linewidth can be quite small, particularly if the upper state is a metastable one, and therefore a single spectral line from some atom might be an excellent monochromatic source if our assumptions of the previous paragraph were true. Unfortunately, in the real world no atom remains free of outside influence. Atoms in a gas are always moving at high speeds, and even those bound in a crystal

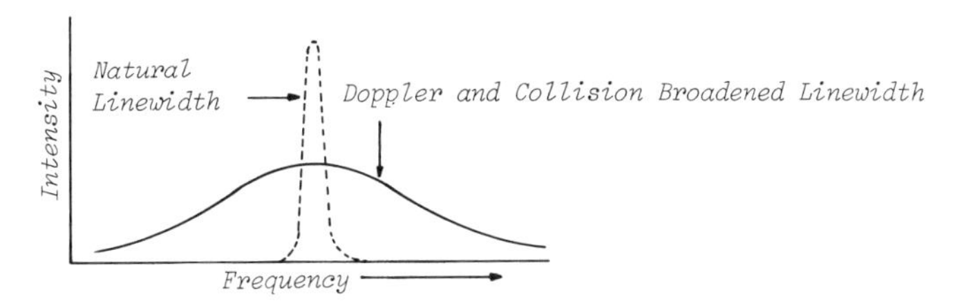

Figure 6–10 Linewidth

such as ruby are vibrating rapidly. Collisions between atoms shorten the excited lifetime and increase the linewidth, a process known as **collision broadening.** The motion of the emitting atoms alone increases the range of frequencies seen by an observer in a process known as **Doppler broadening.** The net result is a linewidth perhaps 10 to 100 times the natural linewidth. This result is illustrated by the graph of Fig. 6–10, which is a plot of the intensity of light emitted at various frequencies versus frequency (the linewidth is indicated by how "fat" the curves are). The broadened linewidth is what one could expect to see from the active medium if no laser action occurred.

The mirror configuration of a laser acts to select only specific wavelengths or frequencies out of the broadened linewidth from the active medium. However, even leaving that effect aside for the moment, we can note that the buildup of stimulated emission in itself tends to narrow the emitted linewidth in an effect known as **spectral narrowing.** As one can see from the broadened linewidth curve of Fig. 6–10, the same power is not spontaneously emitted at all frequencies. Instead, more power is emitted at and near the center frequency and less is emitted at frequencies far from the center. This means that more light near the center frequency will be stimulated and be available to stimulate still more; these frequencies grow in power faster. That is to say, stimulated emission is a kind of cumulative process in which any initial preponderance of light will be magnified. Thus the stimulated emission output from the active medium will have a larger proportion of its light power near the center frequency than the spontaneous emission output would have had, which corresponds to a narrower linewidth.

The optical cavity then uses the wave nature of light to select certain frequencies from the already narrowed linewidth of the stimulated emission. Those wavelengths that just fit into the length of the optical cavity with zeros at each mirror tend to reinforce themselves through constructive interference and build up in amplitude; other wavelengths tend to cancel out through destructive interference after being reflected back and forth many times. Figure 6–11 shows two wavelengths that would be reinforced in the optical cavity. Of course in an actual laser the wavelengths are very much shorter than the length of the optical cavity, which is likely to be 5 to 50 cm. There usually are several preferred wavelengths within the stimulated emission output from the active medium. Then the total laser output becomes a few sharp spikes when plotted versus frequency, as shown in Fig. 6–12. This light has a very small range of frequencies, that is, it is very nearly monochromatic.

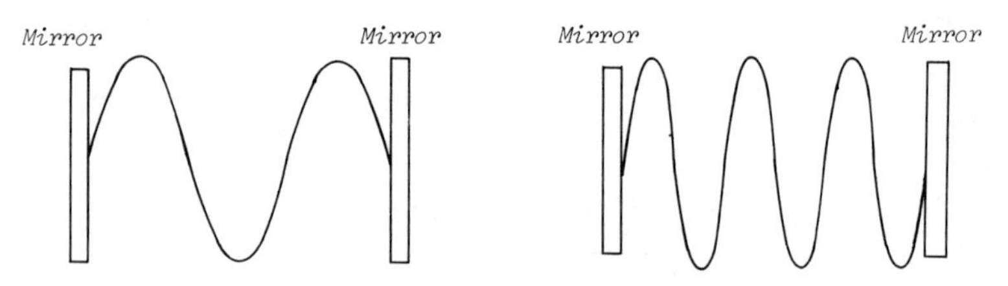

Figure 6–11 Waves selected by optical cavity

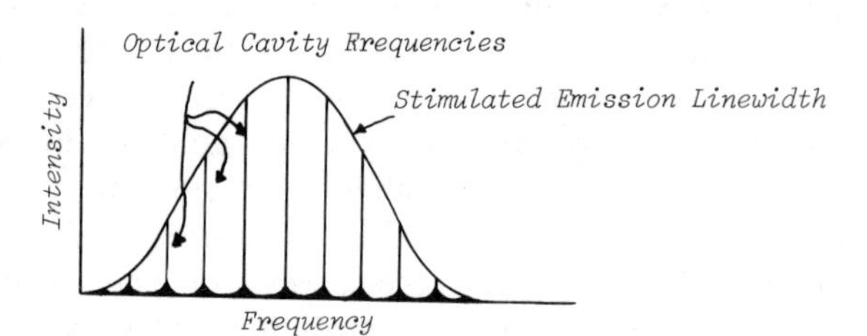

Figure 6–12 Laser output versus frequency

6.7 COHERENCE

Monochromaticity is one aspect of a property of laser light known as **coherence.** In general terms, coherent light is organized in comparison with incoherent light, which is disorganized. There are two kinds of coherence: **spatial coherence,** or coherence in space, and **temporal coherence,** or coherence in time. Monochromatic light is of necessity temporally coherent. Temporal coherence means that there is a fixed phase relationship between two portions of the light emitted at different times that is determined only by the time interval. This is the same thing as saying that the wave trains emitted by a temporally coherent source are very long and unbroken. Then one can be sure that on counting forward or backward a whole number of wavelengths one will find the wave in the same phase, or doing the same thing, as at the starting point. Temporally incoherent light has short wave trains with random intervals beween them, so that moving a whole number of wavelengths along the wave places one at a phase of the wave that cannot be predicted. Figure 6–13 illustrates this property.

The long wave trains of temporally coherent light imply that the source emits without interruption for a relatively long time. The average length of the wave trains emitted is called the **coherence length,** and the average time interval during which the trains are emitted is called the **coherence time.** Assuming that the speed of light when emitted is $c = 3 \times 10^8$ m/s, then with Δl for the coherence length and Δt for the coherence time, we must have

$$\Delta l = c\,\Delta t \tag{13}$$

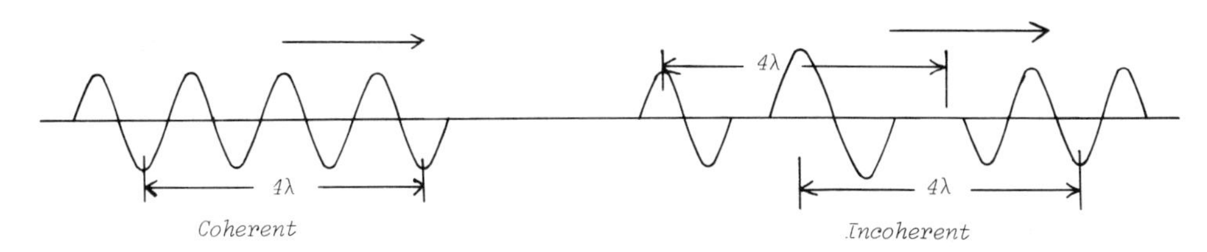

Figure 6–13 Temporal coherence and incoherence

Furthermore, there is always an approximate relation between the time interval during which radiation is emitted by any source and the range of frequencies or bandwidth Δf emitted. We have already noted this relation for the unperturbed spontaneous radiation from an atom in Eq. (12). Here we write the general relation:

$$\Delta f \cong \frac{1}{\Delta t} \tag{14}$$

We may combine the last two equations by reversing Eq. (14) and substituting for Δt in Eq. (13).

$$\Delta t \simeq \frac{1}{\Delta f}$$

$$\Delta l = c\ \Delta t \simeq c\left(\frac{1}{\Delta f}\right)$$

$$\Delta l \simeq \frac{c}{\Delta f} \tag{15}$$

Equation (15) expresses the relation between temporal coherence and monochromaticity in mathematical terms. Strongly monochromatic light has a small bandwidth Δf, which from Eq. (15) implies a large coherence length Δl and correspondingly high temporal coherence.

Example:

What is the coherence length of light from a ruby laser whose output bandwidth is 2×10^9 Hz?

We substitute $\Delta f = 2 \times 10^9$ Hz in Eq. (15)

$$\Delta l = \frac{c}{\Delta f} = \frac{(3 \times 10^8)}{(2 \times 10^9)}$$

$$\Delta l = 1.5 \times 10^{-1}\text{ m} = 15\text{ cm}$$

The bandwidths for ruby lasers are typically on the order of 10^9 Hz, so that corresponding coherence lengths are on the order of centimeters. Typical helium-neon lasers are more monochromatic, with bandwidths about one-tenth those of ruby lasers and coherence lengths of meters.

Spatial coherence, or coherence in space, means that there is a fixed phase relationship between portions of light separated across, rather than along, the beam. This implies that the wavefronts, which are nothing more than lines connecting the same wave phase (usually crests) across the beam, are smooth and predictable for spatially coherent light and randomly bumpy and unpredictable for incoherent light. Figure 6–14 illustrates both cases.

Spatial coherence is necessary for experiments such as Young's double-slit experiment (see Chapter 2). There a portion of the beam of light is being made to interfere with another portion relatively far away (in terms of wavelengths) across the beam. If there were not a fixed phase relation across that distance (if the light were incoherent), then there would not be a constant interference pattern. Instead the interference pattern would change randomly and rapidly, which would cause it to disappear altogether for all practical purposes. Some degree of temporal coherence is needed also, because for off-center parts of the interference pattern a portion of the wave through one slit is interfering with a portion *behind* it through the other slit. However, only coherence lengths equal to a few wavelengths are required, which is not difficult to achieve; the experiment is more likely to be limited by spatial coherence.

An ordinary light source, such as the sun or an incandescent lamp emits incoherent light, that is, a jumble of different wavelengths, short wave trains, and randomly irregular wavefronts. Such light can be made more coherent by passing it through a filter which passes only a small band of wavelengths or frequencies. By reducing Δf this action improves the temporal coherence. The spatial coherence can be improved by passing the light through a small pinhole (often called a **spatial filter**). Because we are then looking at nearly a single point of the wavefront and by Huygens' principle this point acts as a new source of waves, we obtain smooth, spherical wavefronts. However, both types of filter work by throwing away most of the light, and the more they discard the more coherent is the remainder. Before lasers were developed, a high degree of coherence always implied small power or very dim

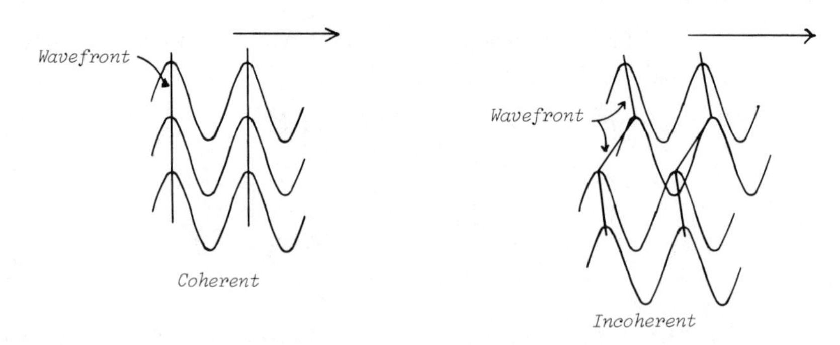

Figure 6–14 Spatial coherence and incoherence

light. But the light from a laser is already very coherent without filtering—stimulated emission and the selectivity of the optical cavity ensure this. And even though the helium-neon power of a few milliwatts may seem low, it really is a tremendous amount to be squeezed into such a small band of frequencies. To draw an analogy from sound, the incoherent light from an ordinary source would be like the noise of a waterfall, while the coherent light of a laser would be like the single, pure tone of a tuning fork.

6.8 APPLICATIONS

The special properties of laser light, such as its collimation, monochromaticity, and coherence, make it quite useful in a variety of scientific and technological applications. One of the simplest and earliest applications uses the parallel beam as an alignment device. Since the beam can be made to spread very little even over long distances, different pieces of equipment can be aligned in a straight line by being put in the beam. Thus, the digging equipment for putting the Bay Area Rapid Transit (BART) tunnel under San Francisco Bay in 1965 was aligned on a helium-neon laser beam. Various other construction and surveying applications are ready-made for a laser beam.

A pulsed laser is a natural candidate for a range-finding device. Very short pulses can be reflected from some target and received back at the position of the laser. The time of flight gives the distance to the target because the speed of light is known. The first astronauts to reach the moon in 1969 left a laser reflector package there for this very purpose. From Lick Observatory in California ruby laser pulses were transmitted through the 120-in. telescope (to improve collimation) and reflected back to the same telescope. The same type of experiment was tried from McDonald Observatory in Texas. The results gave the distance to the moon within 1 m. The beam spread to only about 15 km diameter in the round trip.

This range-finding ability has military applications also. Usually infrared lasers are used for these applications so that the beam is invisible. Other military applications include laser radar (often called **lidar**) and laser target illumination to guide so-called smart weapons. Use of a high-power laser as a weapon itself to burn through objects or people has received widespread speculation. But as yet no such device seems practical, the main problems being the high powers required, the low efficiency of most lasers, and the disruptive effect of the atmosphere. For destructive purposes it would seem that material projectiles are more cost-effective than photons.

The power of many pulsed lasers is high enough to melt or even vaporize most materials at close range if focused. One reason for this ability is the spatial coherence of laser light, which allows it to be focused by a lens down to a spot having dimensions of the order of a wavelength. All the power of the beam is then concentrated into a very tiny area, making it capable of melting almost anything within that area.

This property has been put to use in many materials processing applications. Lasers have been used, for example, to weld and cut metals, cut cloth for clothes, and drill holes in baby bottle nipples.

The high-power laser research sponsored by the military has been turned in recent years toward the more benign use of energy production in the controlled fusion program. It appears possible to trigger the nuclear fusion process by focusing many high-power laser beams from different directions on a small pellet of fuel. A power-producing fusion reactor might use a continuing series of such "mini-H-bombs" to give a controlled, continuous output of power, which could then be used to run an ordinary electric generator. This field of research is very active right now and competing with other methods of obtaining controlled fusion; no one can be sure which method, if any, will turn out to be practical.

This same high-power-density property of focused laser beams has been used in various medical applications. For a condition known as **detached retina,** in which the light-sensitive area at the back of the eyeball peels off, leading to blindness, lasers have been used to make tiny spot welds to hold the retina in place. Here the eye itself does the focusing and the procedure can usually be done on an outpatient basis. Lasers have also been tested for destroying skin cancer cells. They have even been proposed for use in place of the usual dentist's drill to vaporize decayed portions of teeth. Since the decayed area of a tooth should be darker than the healthy area, more of the light energy would be absorbed by the decayed area, aiding the process. Laser scalpels for surgery have also been proposed.

The laser is being used as a diagnostic tool to monitor air pollution. Much information about the types and concentration of pollutants can be obtained by studying the absorption and scattering of laser light as it passes through the atmosphere.

Probably the application with the greatest promise, still largely unfulfilled, is in the area of communications. Almost as soon as the laser was discovered, scientists realized that here was a source of coherent, organized light waves similar to the organized radio waves that are universally used for communications (short-wave, amplitude modulation (AM), frequency modulation (FM), and television broadcast bands). However, the much higher frequencies of light would allow much more information to be carried, at least theoretically. The main stumbling blocks to establishing such laser communications systems have been lack of methods for encoding and decoding information at light frequencies and the disruption of light signals by bad weather. It is just this sensitivity to atmospheric conditions that makes a laser beam a good diagnostic probe, as mentioned above, whereas radio waves are rather insensitive to the same conditions. Still, the dream is coming closer to reality each year as new techniques and equipment are invented. One very promising development is the advent of low-loss fiber optics: thin threads of glass or plastic that transmit light over long distances with astonishingly small loss of intensity. Using these light conductors, a laser communication system would resemble a wired telephone system more than a broadcast radio system but with each optical fiber carrying many times the information possible on the same size wire.

6.8.1 Artistic Use

The application of lasers that is likely to be of greatest interest to readers of this book was probably the one least anticipated by the original scientists and engineers responsible for this new source of light: laser art. The idea of creating artistic designs directly out of light instead of using paints or pigments predates the invention of the laser; Laszlo Moholy-Nagy, an Hungarian artist, was an advocate of such methods as early as 1925.[1] From that time until the development of the laser many artists and groups advanced the same basic concept.[2] But it was the pure light of the laser that really fired artistic imaginations. By the late 1960s a number of artists and designers were using laser light in opera, ballet, and straightforward art exhibits. One of the earliest such exhibits was one called "Laser Light: A New Visual Art" at the Cincinnati Art Museum in December 1969.[3]

One technique used in laser art is to produce a three-dimensional structure of lines of light by multiple reflections of a laser beam from mirrors. Because of the high directionality of laser light, a laser beam generally is not visible from the side: no laser light is traveling at right angles to the beam to enter the observer's eye. On the other hand, a light structure such as the one described above needs to be visible from all angles. This result can be accomplished by filling the air in the volume of the light structure with small scattering particles which will scatter light in all directions, making the beams visible. Water droplets in the form of steam or aerosol spray will work. More commonly, smoke is used for this purpose. If an incense burner is used to produce the smoke, then an olfactory dimension is added to the whole work. Only a small concentration of scattering particles is needed to make the beams visible; if the environment becomes too hazy, the whole volume of the structure becomes filled with diffuse laser light from multiple scattering. An example of one such display, originated by Rockne Krebs, uses a small, darkened room with mirrors covering all four walls above the observer's head (care must always be taken that the direct laser beam does not enter the observer's eye). The laser beam or beams are introduced at an angle through a small hole in one mirror. The multiple reflections off the mirrors form an intricate network of light lines above the viewer's head. Furthermore, two sets of opposite facing mirrors make the network seem to repeat, receding to infinity in all directions. Figure 6–15 illustrates this structure.

A dynamic laser light pattern can be made by having one or more mirrors reflecting the beam move. Moving three-dimensional laser light structures have been produced in acrylic plastic by John Peterson of the University of Cincinnati. Acrylic plastic itself scatters enough light to make the beam visible from the side. One of Peterson's designs used an angled mirror mounted on a clock drive reflecting laser light into the base of a Plexiglas cylinder. This arrangement produced a spiral beam

[1] T. Kallard, *Laser Art & Optical Transforms* (New York: Optosonic Press, 1979), p. 1.

[2] Ibid., p. 2.

[3] Ibid., p. 3.

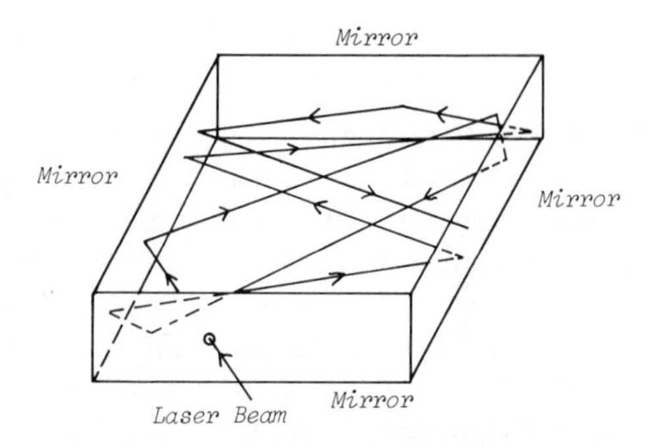

Figure 6–15 Laser light structure

inside the cylinder (kept inside by total internal reflection), which moved around slowly as the mirror moved. This design is shown in Fig. 6–16.

In many cases moving mirrors are used to produce a dynamic two-dimensional light pattern by projecting the laser beam onto a viewing screen. Instantaneously the laser beam produces only a single spot, but if the spot moves fairly rapidly, the persistence of the human eye blends the spot positions into a line. This property of the eye is the same one that allows us to see the separate single frames of a movie as a continuously moving picture. The moving mirror or mirrors can be controlled by hand, by uniformly turning motors, or even by a computer program to produce a stationary or moving pattern of lines. This sort of display has been done on a grand scale with multicolor lasers in laser theater productions such as *Laserium*. It can be done on a much more modest scale by anyone with a helium-neon laser, a few small mirrors, and electric motors. A very interesting random but sound-correlated dynamic pattern can be produced by reflecting the beam from a mirror onto a diaphragm driven by a speaker. Then when various sounds are played over the speaker (for example, music) a changing pattern of light keeps time with the sound. This setup is shown in Fig. 6–17.

One does not have to use smooth reflectors or even reflection at all to produce interesting two-dimensional light patterns. Laser light reflected off various rough surfaces or refracted by somewhat inhomogeneous clear materials produces striking patterns, called **light caustics,** on a screen. Again, the light caustics can be made dynamic by moving the reflecting or refracting object.

Even interference patterns produced when laser light is passed through small

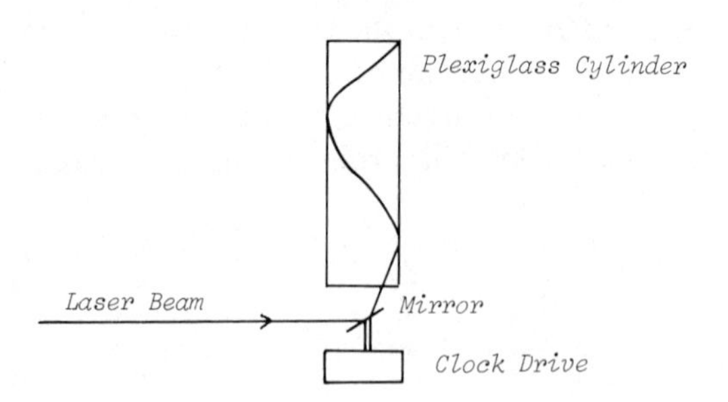

Figure 6–16 Dynamic laser light structure

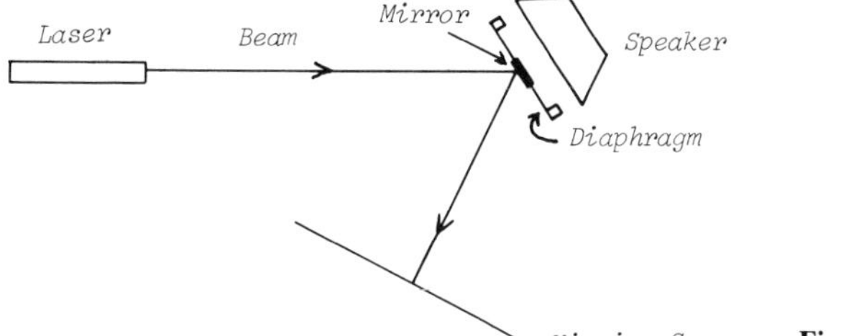

Figure 6–17 Sound-driven laser pattern

apertures of various shapes can be considered as art. Kallard calls these patterns **optical transforms,** and his book shows dozens of these.[4]

Holography, one of the most spectacular and successful of laser applications, has implications for both art and technology. However, this subject is so large that is deserves its own chapter, which follows.

QUESTIONS

1. What does the word *laser* stand for?
2. What is the difference between stimulated emission and spontaneous emission?
3. Who invented the maser? When?
4. Who built the first laser? When? What was the active medium?
5. Who built the first CW laser? When? What was the active medium?
6. Of what are rubies and emeralds composed? Why are rubies red?
7. What is an inverted population? Why is it easier to maintain in a helium-neon laser than in a ruby laser?
8. Why can you not get an inverted population by optical pumping using just two levels in the active material?
9. What type of laser can give several laser wavelengths simultaneously?
10. What would be a typical output power for an ordinary, classroom helium-neon laser?
11. What type of CW laser gives a high-power output?
12. Explain how the end mirors of a laser act to collimate the output beam.
13. What is the coherence length of the light from a helium-neon laser that has bandwidth of 10^8 Hz?
14. A narrow color filter might typically allow through a bandwidth of 1.2×10^{12} Hz out of white light. This corresponds to a wavelength range of over 1 nm at 500 nm. What would be the coherence length of this filtered light?
15. What is the difference between coherent and incoherent light?
16. Describe two different works of laser art.

[4] Ibid., pp. 53–117.

Chapter 7

Holography

Holography is a method of storing images on film which is completely different from ordinary photography. It has been called three-dimensional, lensless photography. The process was invented in 1948 (before lasers) by the British scientist Dennis Gabor (1900–1979) in an effort to improve the resolution of electron micrographs (the pictures from electron microscopes). In 1971 Gabor received the Nobel prize in physics for his discovery. However, the process did not really produce the spectacular results that have since captured the public imagination until 1962, when Emmett Leith and Juris Upatnieks at the University of Michigan first applied laser light and some electrical engineering techniques to holography. Although they were working on optical techniques for displaying radar data, holography since that time has been used in many nontechnical applications such as art and advertising because of its strikingly real and beautiful three-dimensional images.

Holography is basically a two-step process, an exposed film called a **hologram** being produced in the first step and an image being produced from the developed hologram in the second step. The developed hologram does not look anything like the original scene or final image but rather is a complex interference pattern which is unintelligible when viewed in ordinary light. However under the proper viewing conditions the hologram gives rise to a hauntingly realistic image of the original object floating in space.

7.1 ZONE PLATES

To gain an understanding of how holography works, we may start with a much humbler imaging device known as a **Fresnel zone plate,** apparently first constructed in 1871 by the British scientist Lord Rayleigh[1] (1842–1919). This device has imaging properties much like those of a lens, but instead of using refraction it uses interference of light. A Fresnel zone plate is formed by drawing a set of concentric circles that get closer and closer together as their radii get larger (more precisely, the radii are proportioned to the square roots of the integers 1, 2, 3, 4, . . .). Then one must ink in or blacken every other zone, say between circles 1 and 2, circles 3 and 4, etc. Figure 7–1 shows the result. Finally the pattern is photographed and made into a reduced transparency, so that alternate zones are transparent and opaque. Now the finished product will act like a lens with a definite focal length fixed by the size of the smallest circle and the wavelength of light used. That is, a parallel beam of monochromatic light striking one side will be brought to a focus on the other side a distance f away from the plate.

We can see how a zone plate uses interference to produce a focal point by considering Fig. 7–2, which shows plane waves impinging on a Fresnel zone plate. The position of the focal point is such that it is one-half wavelength further away from each successive zone on the average. The focal point is at a minimum distance of f from the first zone, $f + \lambda/2$ from the second zone, $f + \lambda$ from the third zone, $f + 3\lambda/2$ from the fourth zone, etc. If all the zones were transparent, then the waves from successive zones would interfere destructively at the focal point, because the half-wavelength path difference would mean that a crest from one zone always ar-

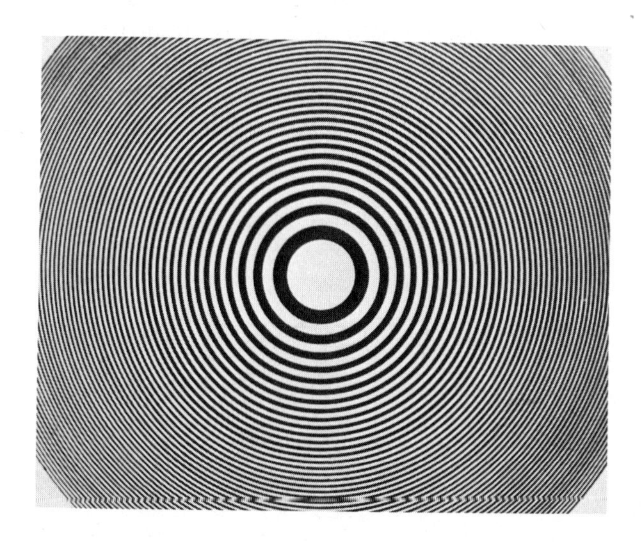

Figure 7–1 Fresnel zone plate

[1] Robert W. Wood, *Physical Optics,* 3rd rev. ed. (New York: Dover Publications, Inc., 1967), p. 37.

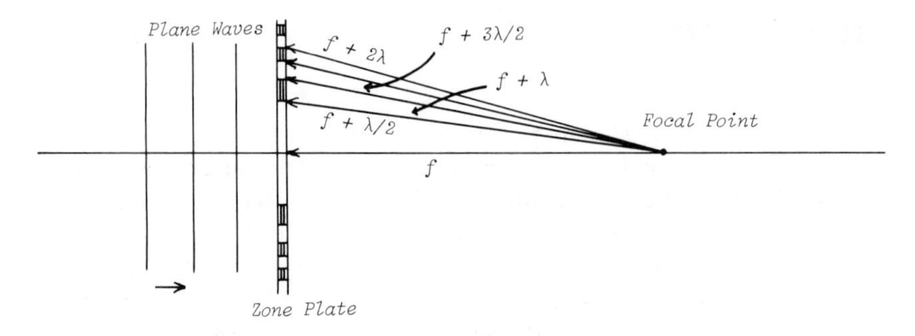

Figure 7–2 Zone plate focusing

rives at the same time as a trough from the adjacent zone. Instead of this situation, however, the zone plate blocks the light from alternate zones, so that the path difference from successive *transparent* zones is one wavelength. Waves from the first, third, fifth, etc. zones arrive at the focal point in phase, giving rise to constructive interference and a bright spot. The zone plate blocks the light that would interfere destructively at the focal point. From these considerations it is obvious that it makes no difference whether even zones or odd zones are opaque; that is, whether the zone plate is a positive or a negative transparency.

An alternative viewpoint is to consider each transparent zone as a source of wavelets according to Huygens' principle, which add up to produce a spherical wave converging to the focal point. There is also another point that fulfills the condition for constructive interference, being at path differences of one wavelength from successive transparent zones; this is the point a distance f away from the other side of the plate. Of course the light, after passing through the plate, is traveling away from this point rather than toward it. Still this point, on the same side of the plate as the approaching plane waves, is a virtual focal point. In other words, the zone plate also gives rise to a set of spherical waves which appear to be diverging from this point. Figure 7–3 shows this effect. In this respect the Fresnel zone plate is unlike a single

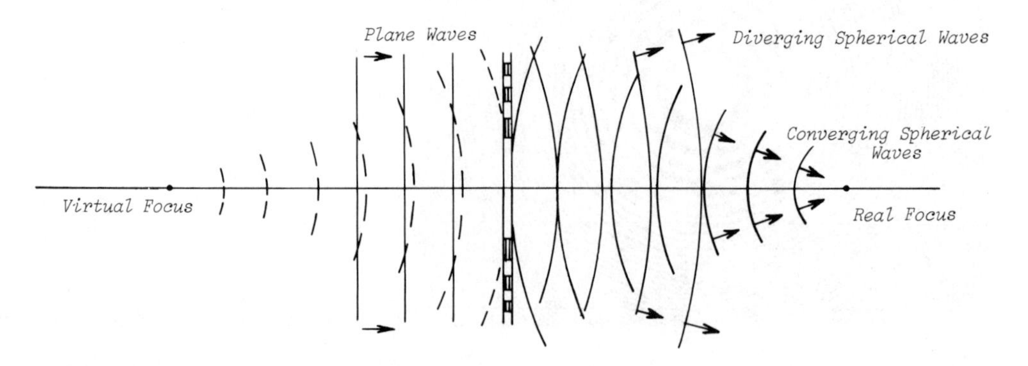

Figure 7–3 Real and virtual foci of zone plate

lens. It appears to be more like two lenses at once, a converging lens and a diverging lens of the same focal length. Similarly, when the incoming light is in the form of spherical waves corresponding to a point object closer than infinity, two images are formed, one real and one virtual, corresponding to the two types of focal point. There is also one additional respect, which will not be important for our purposes, in which a Fresnel zone plate behaves differently from a single lens: it has secondary focal points at $f/3$, $f/5$, $f/7$, etc.

7.2 HOLOGRAM AS A SET OF ZONE PLATES

By this time you might well be asking what all this zone plate business has to do with holography. The answer to that question lies in the fact that a hologram is an interference pattern and the most basic hologram, one of a point object, is an interference pattern which looks almost identical to a Fresnel zone plate. Such a basic hologram can be made by placing a point object in a coherent parallel beam of light (plane waves) and recording on film the interference pattern produced by combination of the plane waves and spherical waves scattered by the point. This arrangement is shown in Fig. 7–4. The resulting interference pattern will be a set of light and dark rings in the same position as those of a Fresnel zone plate with one focal point at the point object. The argument to prove this result is almost identical to the one used to show that the Fresnel zone plate has a focal point at f. The only difference between this hologram and a Fresnel zone plate is that in the hologram the change from a light to dark zone is gradual and continuous rather than abrupt. This basic hologram, illustrated in Fig. 7–5, is often called a **Gabor zone plate,** after the founder of holography.

Essentially the same interference pattern would be formed if the original light were also in the form of spherical waves coming from a point farther away than the object; a Gabor zone plate of somewhat different focal length would result. After all,

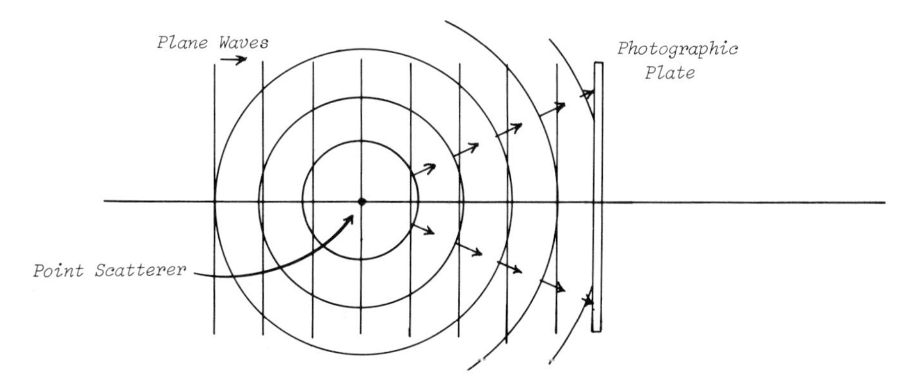

Figure 7–4 Hologram of a point object

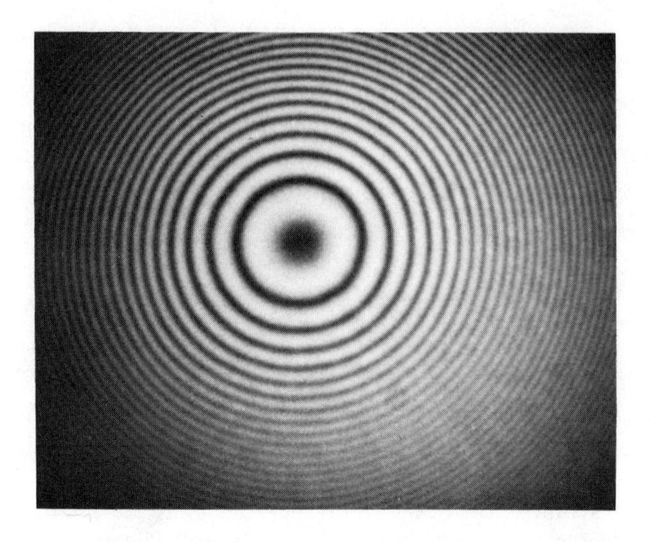

Figure 7–5 Gabor zone plate

even plane waves can be considered as spherical waves originating from a point infinitely far away.

You can see that the hologram of a point object does not in itself look very much like the point object. But a Gabor zone plate has practically the identical imaging properties as a Fresnel zone plate. Therefore, if a developed transparency (positive or negative) of the interference pattern is placed in a parallel beam, it will give a real and a virtual focal point at the same distance as the original object from the plate, which respectively represent a real and a virtual image of the original object. Happily enough, the Gabor zone plate does not have the secondary focal points at $f/3$, $f/5$, etc., which here would just be extraneous and bothersome light. The reconstruction of the point object is shown in Fig. 7–6.

The holographic recording and reconstruction of a point object described above exhibits the general features of holography. The hologram itself is an in-

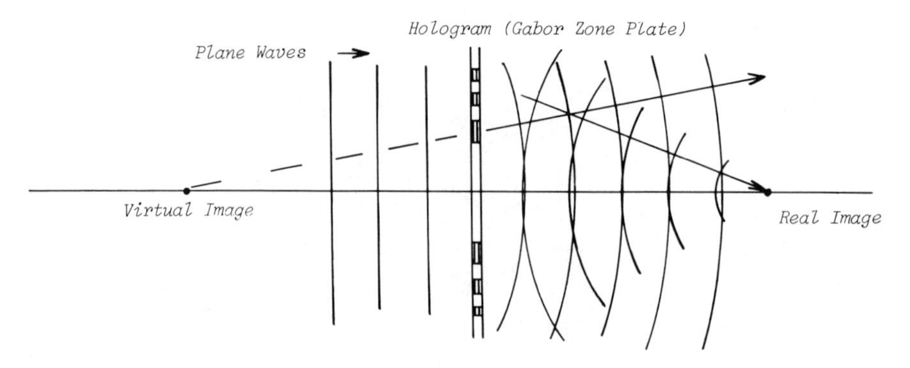

Figure 7–6 Reconstruction of point object

terference pattern between light reflected from the object, called the **object beam,** and light directly from the source, called the **reference beam.** After development, the hologram is put back in its original position in the reference beam with the object removed. Two images of the object now appear: the virtual image at the original position of the object and the real image the same distance away on the opposite side.

For an extended object the principles remain the same, except that in this case each point on the object acts as a source of spherical waves, producing its own Gabor zone plate. The hologram then becomes a complex superposition of Gabor zone plates, each of which will reconstruct one point in the images when illuminated with the reference beam. Although the principles seemed clear, Gabor's early efforts were beset with a number of practical difficulties. First was the lack of a good coherent source of light such as a laser. Instead he used filtered light, which was not very coherent and therefore gave rise to random "noise" in the image. He was also limited to two-dimensional transparencies for objects because, as you can see from Fig. 7-4, a solid, opaque object would scatter light back toward the source rather than toward the photographic plate as required. Finally, in the reconstruction process the observer generally ends up viewing light from both the real and virtual images together (see Fig. 7-6), so one image becomes unwanted noise in the other.

7.3 IMPROVED HOLOGRAMS

Leith and Upatnieks introduced to holography both the laser, with its superior coherence, and an off-axis technique that allowed three-dimensional objects and visual separation of the real and virtual images. By splitting the beam from the source and sending one part to be reflected from the object while sending the other part directly to the photographic plate, they could bring in the object and reference beams from somewhat different directions. Figure 7-7 shows two of the many ways this result can be accomplished.

In the off-axis method the hologram is still a superposition of Gabor zone

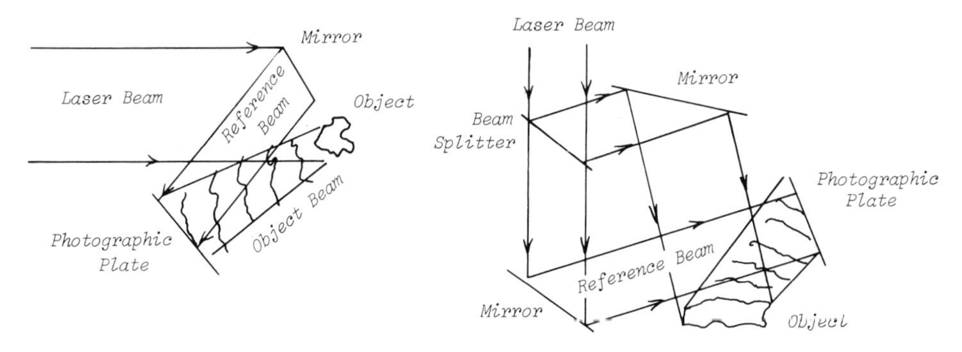

Figure 7-7 Off-axis holographic techniques

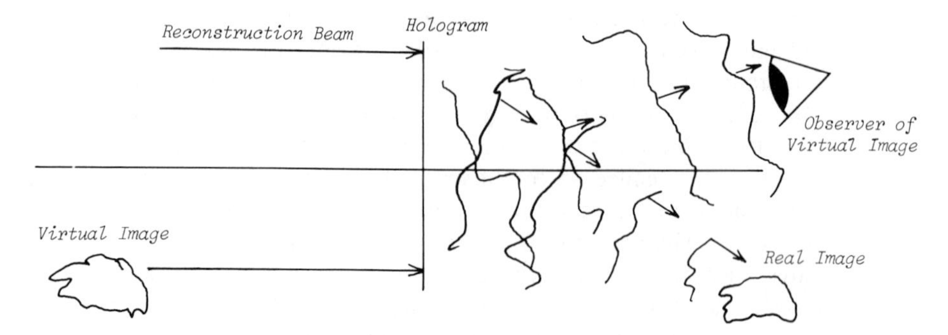

Figure 7–8 Reconstruction with off-axis hologram

plates, one from each point on the object, but now none of the zone plate centers are on the plate. Only the outer edges of the zone plates are recorded, leading to very fine fringes visible only under a microscope. Correspondingly high-resolution film must be used to accurately record the pattern. Even though only the outer portions of the Gabor zone plates appear in the hologram, each one still accurately reproduces one point in the image; consideration of Fig. 7–2 and the ensuing explanation of zone plate focusing should convince you that any portion of the zone plate will produce a point image. Reconstruction is again accomplished by placing the developed hologram in the reference beam alone, as shown in Fig. 7–8, from which it is clear that the real and virtual image are seen by viewing from different directions, so that their light does not become mixed together.

7.4 HOLOGRAPHY AND PHOTOGRAPHY

It is of interest at this point to compare holography with ordinary photography, for both are methods for storing and displaying images. It is clear that a hologram contains much more information than a photograph taken of the same scene. Since the Gabor zone plates of the hologram give each point of the image at its proper distance, a strikingly realistic three-dimensional image results. The virtual image, which is easier to view, looks just like the original object in its original position. It is so realistic that the eye (or a camera) must change focus to see its different parts clearly, just as if the observer were viewing the object. Furthermore, the observer can obtain the appropriate changes in perspective by a slight head movement, actually looking around foreground objects in the holographic image. Such effects are impossible with a photograph, which gives us a two-dimensional image from just one viewpoint. Even a stereo photograph in a stereo viewer, although it reproduces dual viewpoints associated with the two eyes, only records the scene from a fixed head position. One hologram is equivalent to many photographs all taken from different angles.

Considering the virtual image from a hologram, it is as if the hologram had somehow "frozen" the light waves from the original scene and the reconstruction process then released them to continue on their way exactly as they would have if the photographic plate had not stopped them in the recording process. In fact, Gabor originally called his method **wavefront reconstruction.** A photograph cannot possibly give this kind of reconstruction because it only records information about the wave amplitude at the plane of the film. A hologram records both amplitude and phase information from the object waves because it records the interference pattern they make with the reference waves.

Another way in which a hologram differs from a photograph is in the rather amazing fact that any piece of a hologram is itself another hologram of the entire scene! The information from each point in the object scene is spread across the entire hologram. That is to say, each Gabor zone plate (corresponding to one object point) is recorded over the whole film. Use of a small portion of the hologram may degrade the image and restrict the viewpoints from which it may be seen; however, the whole scene will be present in the image. A photograph, on the other hand, localizes the information from one object point at one point on the film; if that portion of the film is lost all the information about that part of the scene is also lost.

Of course we do not get something for nothing. A price is exacted for all that extra information in a hologram—a hologram is much harder to make than a photograph. Coherent light sources, special high-resolution film, and isolation from mechanical vibration are required. The latter requirement is necessary because the hologram is really an interference pattern, which can be destroyed by any movement of the object as small as a fraction of a wavelength. Special supporting systems on air or sand are often used to limit vibration of all components during exposure of the film. Alternatively, a pulsed laser can be used to give an exposure so short that nothing has time to move.

The requirement for a coherent source also stems from the fact that a hologram is an interference pattern. The source of light needs to be spatially coherent so that the wavefronts of the reference beam are regular over the film plane; any irregularities at the film plane should only carry information about the object, not random noise from the source. Temporal coherence is necessary because light from a point on the object may travel a longer or shorter distance to the film plane than the reference light with which it must interfere. This path difference means that some light in the original beam from the source must interfere with other light from the beam which is many wavelengths ahead of or behind it. No stationary interference pattern will result unless the coherence length is sufficiently large.

7.5 HOLOGRAMS AS SETS OF HYPERBOLIC MIRRORS

So far the holograms we have considered have been two-dimensional themselves—they could be recorded on very thin photographic emulsions. We can see additional capabilities of holography if we consider the interference pattern in three

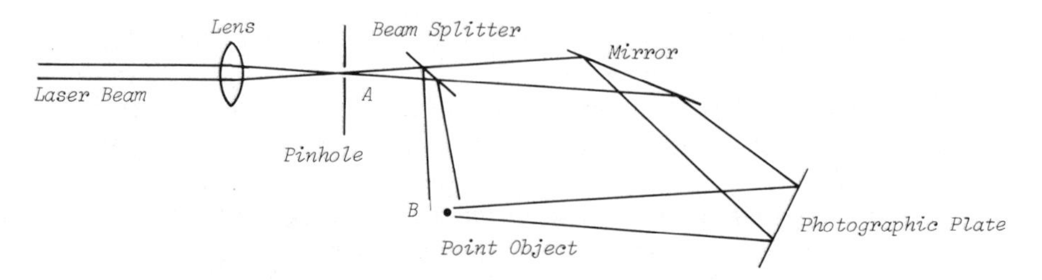

Figure 7–9 Holography with spherical reference waves

dimensions, recorded in a thick photographic emulsion. In this case an ingenious geometrical optics approach[2] allows us to predict many interesting imaging properties of the thick hologram. Again we start with the most basic hologram of a point object. Also, we shall consider the reference beam to consist of spherical waves coming from a point physically separated from the point object. This is usually the case in reality since the laser beam is commonly focused down to pass through a pinhole, in a process called **spatial filtering,** to produce a larger spatially coherent beam. Then the pinhole A becomes the center of curvature for the spherical reference waves. Figure 7–9 illustrates this situation.

Now the interference pattern between the reference beam and object beam is simply that of two point sources. This pattern is formed by two sets of spheres, with each set centered at a different point (two sets of circles in the plane of the paper). This pattern is shown in Fig. 7–10. This situation is also identical to the double-slit interference pattern shown in Chapter 2 (see Fig. 2–8). It can be shown that construc-

Figure 7–10 Two sets of overlapping circles

[2]Tung H. Jeong, *Am. J. Phys.*, 43 (1975) 714.

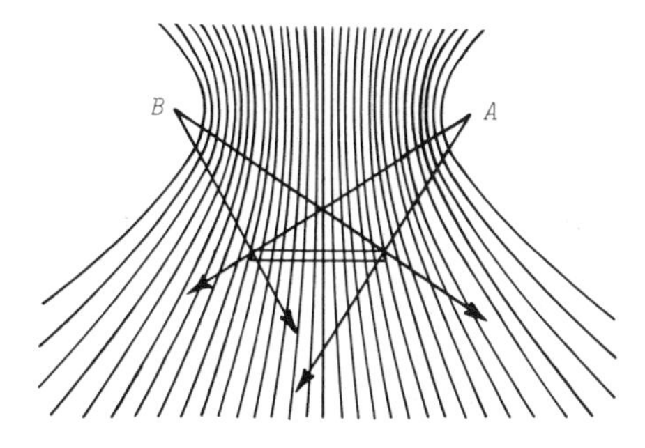

Figure 7-11 Hologram recording of hyperboloids

tive interference occurs in three dimensions along surfaces that are hyperboloids of revolution. In two dimensions in the plane of the paper, constructive interference occurs along the lines connecting the points of intersection of the circles, which are hyperbolas. These hyperbolas, shown in Fig. 7-10, are just slices of the hyperboloids, which can be visualized by rotating Fig. 7-10 about an axis through the two point sources. Furthermore, the foci of all the hyperbolas (hyperboloids in three dimensions) coincide at the two point sources. We have a family of hyperbolas with the same foci, like those of Chapter 4 (see Fig. 4-45).

In the holography process part of this family of hyperboloids is recorded as the hologram, as shown in Fig. 7-11. Note that now we are assuming the pattern to be recorded in depth through a thick emulsion. The pattern produced at the plane of the face of the photographic plate would still be a portion of a Gabor zone plate. When the hologram is developed, each of the bright lines in the emulsion will become partially reflecting and therefore can be considered as a portion of a hyperbolic mirror. It is this model of a thick hologram as sets of hyperbolic mirrors that we study here.

The virtual image of a hologram can be seen by placing it in the original reference beam and looking through the hologram back toward the original position of the object. However, we know that the two foci of a hyperbolic mirror are virtual images of each other. Therefore every reflecting portion in the hologram reflects light from point A as if it had come from B, as shown in Fig. 7-12. The virtual image appears at point B, the original position of the object that produced the hologram. For an extended three-dimensional object, we must assume that each point on its surface

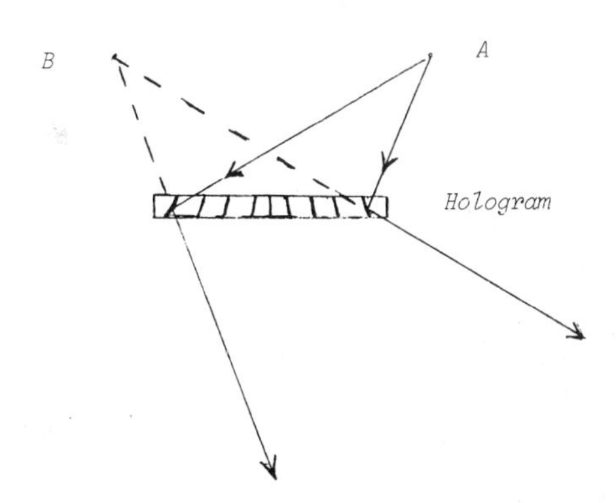

Figure 7-12 Reconstruction of virtual image

acts like a point object, so that the whole object gives rise to many sets of hyperbolic mirrors in the hologram. Each point of the object produces one set of hyperbolic mirrors with the same foci, and each of these sets of mirrors produces one point in the image during reconstruction.

The real image can be produced by a beam converging toward *A* but striking the hologram first. Such light would be brought to a focus at *B* after reflection (see Fig. 4-42). This result is shown in Fig. 7-13.

Of course we know from our earlier discussion of zone plates that interference of light must play some part in the imaging properties of a hologram; focusing by reflection from hyperbolic mirrors cannot be the whole story. As a matter of fact, the sections of hyperbolic mirrors in the hologram are so small and so poorly reflecting that ordinarily we would expect to see no image. It is only when the reconstructing beam strikes the hologram at precisely the same angle as the original reference beam that we get a bright image, because at that angle constructive interference takes place. When the reconstructing beam strikes at this particular angle, the path difference for light reflected from successive hyperbolic mirrors in the hologram is exactly one wavelength because of the way the original interference pattern was formed. Thus, constructive interference between the light from successive hyperbolas greatly brightens the image at this one angle of illumination; at angles significantly different, the image cannot be seen, although just the geometrical optics treatment of hyperbolic mirrors would predict about equal image brightness at other angles. This property of holograms allows one to record and reconstruct completely independent scenes on the same plate. One scene is recorded with one reference beam angle and the next scene with a slightly different angle. After development, the different scenes can be reconstructed by varying the angle of the reconstructing beam.

This constructive interference from successive reflecting surfaces can even serve to select the correct reconstruction wavelength from among many in the reconstruction light. To accomplish this selection, however, a great number of reflecting surfaces in the emulsion is required. If enough reflecting surfaces were present in the hologram, then white light could be used in the reconstruction, which greatly reduces the cost for display of holography and increases the art and advertising applications. By 1966 several experimenters had produced such white light holograms. The tech-

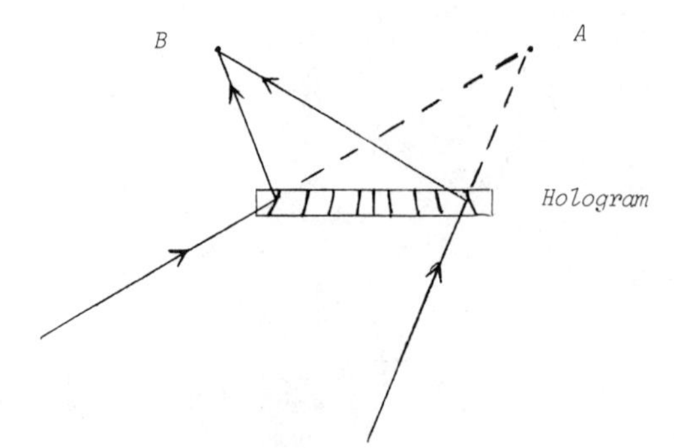

Figure 7-13 Reconstruction of real image

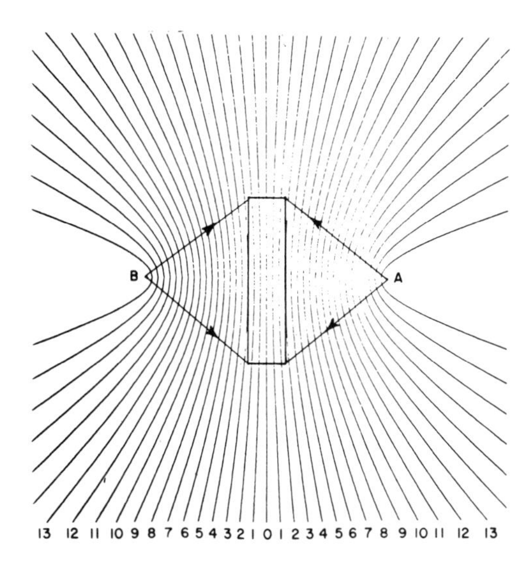

Figure 7–14 Production of white light reflection hologram

nique for getting a maximum number of reflecting planes in the emulsion consists of positioning the photographic plate so that the object beam and reference beam approach from opposite sides. This is illustrated by Fig. 7–14. The recording still must be done with a laser; only the display phase uses white light. If the film remained undistorted in processing, constructive interference between light waves reflected from different surfaces in the hologram would select exactly that laser wavelength out of the white reconstruction light. In practice, however, the emulsion usually shrinks somewhat during processing, and the reduced spacing between reflecting surfaces selects a wavelength somewhat shorter than the one used for recording. Because of the geometry shown in Fig. 7–14, in a white light hologram the reconstruction beam and the light reflected off the hyperbolic surfaces are essentially in opposite directions. The viewer must look in from the same side as the reconstruction beam. Often a dark backing is put on the hologram to cut out disturbing transmitted light. These are called **white light reflection holograms.**

7.6 NEW TYPES OF HOLOGRAMS

White light transmission holograms are also possible through a technique invented by Stephen Benton at the Polaroid Corporation in 1968. To make this type of hologram, a more complex setup is required to record a horizontal slit aperture in front of the object and to place a real image of the object near the plane of the hologram. The horizontal slit restricts the angle in the vertical direction from which the final image can be viewed; in other words, there will be no vertical parallax. However, since we normally view a scene with our two eyes separated in the horizontal direction, the loss of vertical parallax will not seriously affect the three-dimensional appearance of the

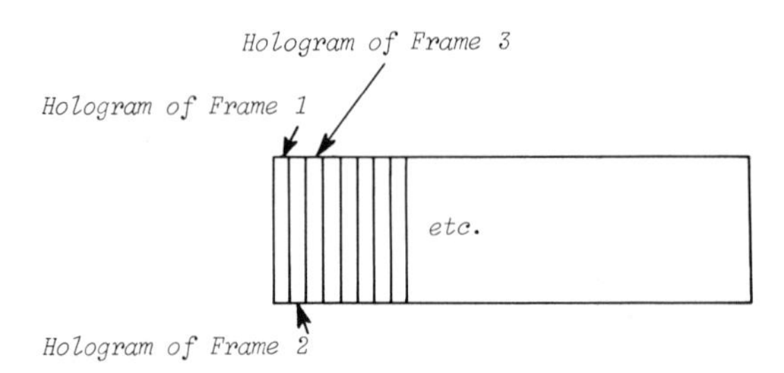

Figure 7-15 Multiplex hologram

final image. Now if white light is used in reconstruction, the various spectral color images will be projected out at different vertical angles so that an observer can see different color images by moving his or her head up or down. Because of this multicolor effect, these are known as **rainbow holograms.** Essentially, in this case one has traded different vertical views of the scene for different colors of one vertical view with white reconstructing light.

One of the most striking types of holograms was developed about 1972 by Lloyd Cross. Called a **multiplex hologram,** it can display a limited amount of motion as well as depth. Here the whole scene is first recorded on ordinary motion picture film in ordinary light with a movie camera, but the scene must be on a rotating turntable or the camera must slowly circle the scene. Then, in the next step, a special device converts the developed frames of the movie into strip holograms. That is, by using laser light and one movie frame as an object, a hologram is made as a vertical strip on a holographic film. Next to this strip, a strip hologram of the next frame is made, until finally all the movie frames of the scene are recorded in vertical strips on one long holographic film, as indicated in Fig. 7-15.

To display such a hologram, it is mounted on a clear cylindrical surface and illuminated by a white light source at the cylinder axis. The two eyes of a viewer looking into the cylinder see different holograms of different movie frames, which will give slightly different views of the scene because of the relative rotation between the movie camera and the scene. These different viewpoints will give rise to the normal stereo three-dimensional appearance, but furthermore, as the observer moves around the cylinder, his or her eyes will see other movie frames, and the image will reproduce the motions recorded by the motion picture camera. Again in this case, vertical parallax is lost because there is no depth information in a single movie frame and the camera-scene rotation that gave such information was only a horizontal motion. For the present, such displays are limited to about 45 s of motion.

7.7 APPLICATIONS

The applications of holography are still being discovered. Gabor's original idea was to improve the images obtained with the electron microscope. This application illustrates the important fact that the hologram may be made with one wavelength and the image may be reconstructed with another. Even if the original wave is one which does not expose photographic film, the information can often be transferred to film by some electronic devices so that the final image can be displayed visually. Leith and Upatnieks were using such techniques to display radar information as a visual image. In this case the radar waves produced the information in the hologram, but light was used for reconstruction. Furthermore, the final image size may be enlarged or reduced as compared with the original scene by adjusting the radius of curvature of the reconstructing beam with respect to that of the original reference beam. In particular, if both such beams are plane waves, then the magnification is unity[3] (the image is the same size as the original scene). It has been suggested that holography with ultrasonic waves could replace diagnostic x-rays, giving better images at reduced risk to the patient.

In some types of scientific research holography could be very much more useful than photography. For example, if all the positions and appearances of objects in some volume of space were to be recorded simultaneously, a photograph would have depth-of-field problems, as only one plane in the volume would be in precise focus in one photograph. However, a hologram taken with the very short pulse of a pulsed laser would record all the information from the illuminated volume at once. During the reconstruction process the researcher could focus attention on any desired details in the volume. Such capability would be important in studying a large volume such as a suspension of particles in air, or even a small volume such as a living cell under the microscope, which gives a very small depth of field.

The hologram is considered to be a possible type of optical memory. A great deal of information can be stored in a small space on holograms, particularly if many images are stored at different angles on the same film, as described earlier. Such optical memories could be parts of optical computers in which the signals are carried on light beams instead of wires.

Another technological application of holography is in the field of interferometry. If a hologram is made and placed back in the system in the reference beam with the object still present, the light from the actual object and its virtual image superimposed on it will interfere. Even the tiniest deformation of the object from its shape when the hologram was taken will show up in the interference fringes. Thus the effect of various stresses on the object can be analyzed.

However, the most immediate and obvious application, and the one most likely to be of greatest interest to the reader, is in the field of art and advertising. Just the unique beauty and realism of the holographic image recommend the technique for these purposes. Artists were quick to adopt the new process. The first holographic art exhibit was in 1970 at the Cranbrook Academy of Art in Bloomfield Hills, Michigan.[4] This show featured 27 holograms by various artists. It was followed by "N-Dimensional Space," a 1972 exhibit at Finch College Museum of Art;

[3] John B. De Velis and George O. Reynolds, *Theory and Applications of Holography* (Reading, Mass.: Addison-Wesley Publishing Co., Inc., 1967), p. 68.

[4] T. Kallard, *Laser Art & Optical Transforms* (New York: Optosonic Press, 1979), p. 7.

"Holography 75: The First Decade," in 1975 at the International Center of Photography in New York; "Holography: The 3-Dimensional Medium," in 1976 in Stockholm, Sweden; and others. By 1978, an exhibition of holography in Tokyo featured 90 holograms and was viewed by more than 50,000 people in 2 weeks.[4] Holography appears certain to become a permanent part of the American art scene since the 1976 founding of the Museum of Holography by Rosemary Jackson in New York City.

QUESTIONS

1. When and by whom was holography invented?
2. Who was the first to apply lasers to holography?
3. How do you make a Fresnel zone plate? Who was the first to do so?
4. How does a Fresnel zone plate focus an impinging plane wave?
5. In what ways is a Fresnel zone plate like a lens and in what ways is it different?
6. What is a Gabor zone plate and how is it related to a Fresnel zone plate and to holography?
7. A hologram of an extended object can be considered as a superposition of what?
8. Describe how to make an ordinary off-axis hologram.
9. If a hologram is broken in half and only one half is put in the reconstructing beam, what do you see?
10. How is a hologram different from a photograph?
11. What kind of figures are formed by the surfaces of constructive interference inside a thick hologram?
12. Describe how to make a white light reflection hologram.
13. What kind of information is missing in a white light transmission (or rainbow) hologram?

Part III

Vision

Chapter 8

The Eye

Human vision is a complex and wonderful process that is not completely understood even today. It involves the eye, the nerves, and, perhaps most importantly, the brain. Vision starts with the eye, so that is where we start out study: More is known about this first component of the visual system than about the other parts, and as we delve more deeply, the presentation will of necessity become more sketchy. Figure 8–1 shows a horizontal cross section of the right eye of a human being. It will be convenient to consider the various structures within the eye in the same order in which they are encountered by light entering the eye.

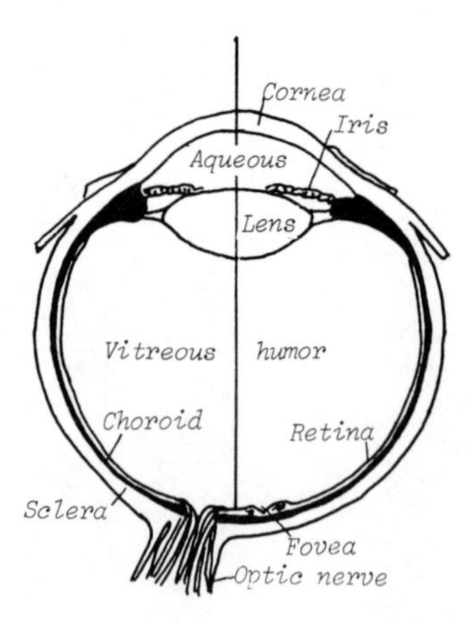

Figure 8–1 Horizontal cross section of right eye

8.1 OUTER SHELL

The outer portion of the eyeball is a rigid, white spherical shell called the **sclera,** about 22 mm in diameter and 1 mm thick. The sclera gives the eye its structural integrity and is what you can see as the "white of the eye." However, the sclera itself is opaque and so to admit light it merges at the front with a smaller, transparent spherical shell known as the **cornea.** If the eyes are the windows of the soul, the cornea is the window of the eye. The cornea has a radius of curvature of about 8 mm and a refractive index $n \cong 1.37$[1] (note the more pronounced corneal curvature in Fig. 8–1). This structure is more than just a window, however: it is the first and major refractive element in the eye, causing most of the convergence of light rays toward the real image that will be formed inside the eye.

Immediately behind the cornea is a space filled with a watery liquid called the **aqueous humor.** Optically the liquid is just water, with $n = 1.33$, but the aqueous humor does have a definite purpose other than merely filling up space. The cornea is an unusual body tissue in that it is not supplied with blood vessels. If it were, they would disrupt the optical clarity needed. Instead, the cornea is supplied with nutrients by the aqueous humor, which is totally renewed about once every 4 h.[2] It is worth noting that the arrangement described so far makes the cornea a somewhat different kind of lens than those discussed in Chapter 4 because it has a different medium on each side of it, air in front and water behind. Because there is little change in refractive index when the light passes from cornea to aqueous humor, there is little additional bending at that boundary.

8.2 PUPIL

At the back of the chamber containing the aqueous humor is the colored **iris** of the eye. The iris merges with pigmented connective tissue lining the interior of the sclera and called the **choroid.** The opening in the center of the iris, through which light passes, is the **pupil** of the eye. The iris contains delicate muscles, which can change the pupil size in response to the amount of light entering the eye. The pupil diameter ranges from about 2 mm on a bright day to 8 mm under very dark conditions, and the adjustment takes place automatically within tenths of a second.[3] This pupil response is shown in Fig. 8–2, which gives pupil diameter averaged over many subjects versus light level. The round shape of the human pupil is not the rule within the animal

[1] *The Science of Color,* (Washington, D.C.: Optical Society of America, 1963), p. 74.

[2] R. L. Gregory, *Eye and Brain: The Psychology of Seeing,* 3rd ed. (New York: McGraw-Hill Book Company, 1978), p. 49.

[3] Albert Rose, *Vision: Human and Electronic,* (New York: Plenum Press, 1973).

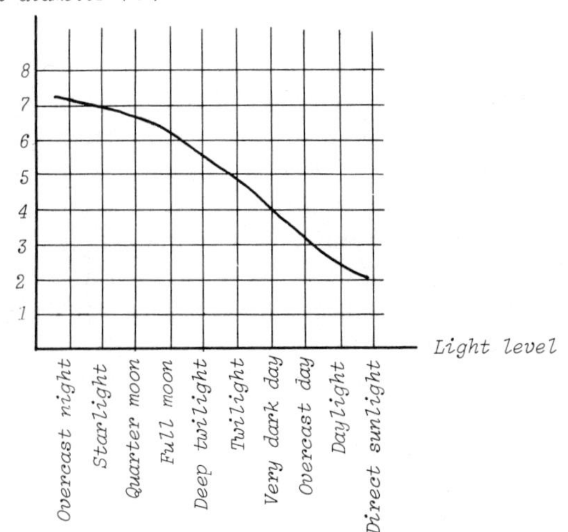

Figure 8–2 Average pupil diameter versus light level

kingdom: some animals have vertical slits (snakes, alligators, cats) while others have horizontal slits (goats, horses). The pupils of both eyes should open and close together even if only one eye is stimulated; this coordinated response is called the **consensual pupillary reflex.** Thus, if in a darkened room a penlight is shined into one eye, both pupils will constrict. Failure of this reflex could indicate brain damage.

Although pupil size changes in response to lighting conditions, that does not mean that other factors do not play a part. Various drugs can affect pupil size. More interestingly, our attitude toward the scene observed also has an influence. The pupils tend to open or dilate when one views a scene of great interest, as when male subjects view photographs of women or female subjects view photographs of men: there can be as much as a 30 percent increase in pupil diameter with no change in light level.[4] The reverse is also true, in that the pupils constrict when the observer views a distasteful scene. It is almost as if the eye-brain system is trying to take in more of an appealing scene and reject an unappealing one, all in a totally subconscious mode. One interesting sidelight of this type of research is that male subjects found that a photograph of a female face with the pupils retouched to appear larger was more appealing than the unretouched photograph, even though none consciously noticed the pupil difference. Apparently we may often read pupil size subconsciously as a clue to people's reaction to us; if we appeal to someone, that person's pupils should be relatively large when he or she is looking at us, and in a reinforcing manner that person consequently impresses us as warmer and more appealing. Love at first sight? Conversely, small pinpoint pupils will give a face a cold, hateful look. In passing we might also note that close, fine detail work will also cause the pupils to constrict, apparently to reduce aberrations in the final image.

[4] Eckhard H. Hess, "Attitude and Pupil Size," *Scientific American,* April 1965.

8.3 LENS

Immediately upon passing through the pupil, light enters the **lens** of the eye (sometimes called the **crystalline lens**). This clear, lens-shaped piece of tissue has a refractive index ranging from 1.40 at its center to 1.38 in the outer portions.[5] This is the second major refractive element in the eye, but it still does not do as much of the focusing as the cornea because its index of refraction is not greatly different from those of the media on either side of it. Instead, the major function of the lens is to give the fine adjustment in focusing necessary to sharply image objects at different distances from the eye. This adjustment for different object distances is known as **accommodation,** and is accomplished by an actual change in shape of the lens caused by the action of the muscles from which it is suspended. When the muscles are relaxed, the lens assumes its flattest, least curved shaped, and the eye is focused for distant objects. When the muscles are tensed, the lens is squeezed into a more curved shape, and the eye is focused on nearby objects. Accommodation, like the pupillary reflex, takes place automatically, but it is possible for a person to consciously override this autofocus feature of the eye and thereby defocus an object being viewed. Because close work requires continuous muscle tension within the eye, it can lead to a feeling of eyestrain; gazing at a distant (15 to 20 ft or farther) scene allows relaxation of those muscles.

The closest object distance upon which the eye can focus is called the **near point** of the eye. This distance generally increases with age because of the loss of elasticity of the lens and weakening of the supporting muscles. Table 8–1 gives average values of the near point versus age, documenting this effect, known as **presbyopia.** Often 25 cm is taken as a standard value. Note that it is normal for a person's near point to surpass the length of his arms somewhere between the ages of 50 and 60. Then eyeglasses are called for because reading material can no longer be held far enough away to be in focus. At the other extreme, infants probably have such a close near point that objects are still in focus even when they are brought so close that they cause the eyes to cross (a familiar action by babies examining their hands or a toy).

TABLE 8–1 AVERAGE NEAR POINT AT VARIOUS AGES

Age (years)	Near point (cm)
10	7
20	10
30	14
40	22
50	40
60	200

[5] *The Science of Color,* p. 74.

One serious defect of the lens is **cataract.** This vision-threatening loss of transparency in the lens is most common in old age; its cause is unknown. Treatment consists of surgical removal of the clouded lens and is one of the most frequent types of operation (and one of the most successful). Formerly, after such cataract surgery thick spectacles were required for clear vision. More recently contact lens and even artificial lens implants have become increasingly common, usually with much more satisfactory results for the patient.

Behind the lens the inner chamber of the eye is filled with the jellylike **vitreous humor,** which has a refractive index again very close to that of water. The vitreous humor is essentially structureless but may have bits of cellular debris in it that give rise to faint shadows in the field of view known as **floaters.**

8.4 RETINA

Within the choroid, covering the posterior two-thirds of the eyeball, is the final destination of the light, the light-sensitive layer called the **retina.** The purpose of everything in front of the retina is ultimately to form an appropriate real image of the outside world on this layer. This image, like that of a single thin lens, is inverted on the retina. The effective focal length of the relaxed eye is about 17 mm, which means that the retinal image is the same size as that which would be formed by a single 17 mm focal length lens. When the eye is fully accommodated for nearby objects, the effective focal length shrinks to about 14 mm.

Often when people first learn of the inverted retinal image, they are puzzled as to why we do not see the world upside down. However, a little contemplation will clear up the mystery. Seeing is emphatically not just the passive transfer of the retinal image to some portion of the brain for inspection by some kind of little demon there (after all, how would the demon "see" it?). The retinal image has always been inverted; usually we have had no experience of anything else. The conscious "picture" we see is simply the way the brain interprets the retinal image as it exists. Along these lines, some very interesting experiments have been done with human subjects wearing inverting goggles. The American psychologist G. M. Stratton actually wore an image-inverting device for 8 days as long ago as the 1890s. As far as we know he was the first person to see with a noninverted, or right-side-up, retinal image. Of course, ordinary actions were difficult at first, but by the fifth day things seemed almost normal although his view of parts of his own body still seemed confused.[6] When he removed the device after 8 days the normal view of the world with inverted retinal image seemed bewildering but not upside down; the strangeness lasted only a few hours. Apparently, the visual system can learn to function with very different kinds of retinal images, even those distorted in more complex ways.[7]

[6] Gregory, *Eye and Brain,* p. 205.

[7] Ivo Kohler, "Experiments with Goggles," *Scientific American,* May 1962.

8.4.1 Nearsightedness and Farsightedness

Two common visual defects result from failure of the eye to focus images precisely at the position of the retina. **Nearsightedness,** or **myopia,** occurs when the eyeball is longer than normal so that the relaxed eye forms the image of a distant object in front of the retina. **Farsightedness,** or **hyperopia,** occurs when the eyeball is shorter than normal so that the relaxed eye forms the image of a distant object behind the retina. Figure 8–3 illustrates these conditions. Sometimes the word *hypermetropia* is used in place of hyperopia. Actually the nearsighted person can see nearby things fairly clearly; we could say the myopic person's far point (the farthest away one can focus) is less than infinity. You can understand this result if you remember our earlier discussion of the image-forming properties of lenses; as an object is moved closer to a lens, its real image moves farther away. An object close enough to the relaxed myopic eye will produce an image on the retina, and any object closer can then be focused by accommodation. The opposite is true for the hyperopic person: distant objects can be seen clearly, but the near point is abnormally far away. In this respect Fig. 8–3 is somewhat misleading because the eye is shown relaxed, whereas the hyperopic person would actually accomodate slightly (fatten the lens) to bring the image onto the retina. Only in this case, because of the shortness of the eyeball, full accommodation will bring into focus only things relatively far away; any closer object will be blurred. Note that both the presbyopic and hyperopic eye are characterized by abnormally distant near points and could be called farsighted, but for different reasons, the former because of the lens, the latter because of the shape of the eyeball. The nearsighted eye can be corrected by a diverging lens and the farsighted eye by a converging lens, as shown in Fig. 8–4.

8.4.2 Fovea and Blind Spot

Structure within the retina itself can also be seen in Fig. 8–1. Almost on the axis of the eye in the retina is the area of greatest visual acuity, called the **fovea.** That means that any portion of the real image formed in the eye that falls on the fovea (and only that portion) will be seen clearly and sharply. When we "look at" something we are putting its image on the fovea. The fovea is quite small, being only about 0.25 mm in diameter and representing a tiny portion of the field of view, equivalent roughly only

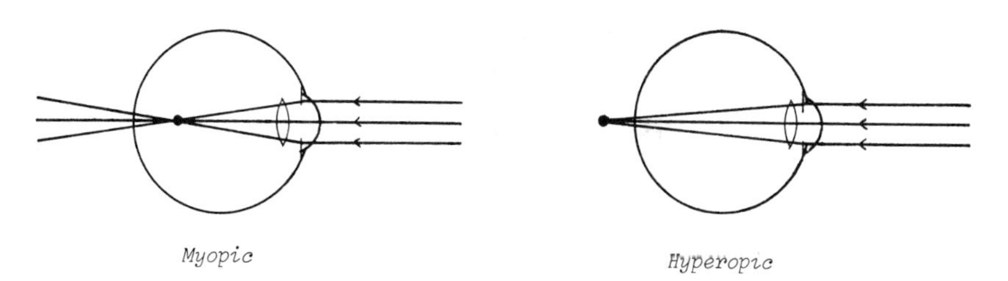

Figure 8–3 Myopic and hyperopic eyes

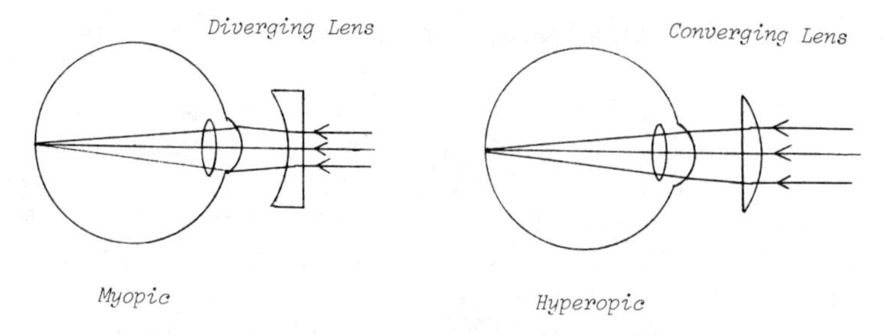

Figure 8–4 Correction of myopic and hyperopic eyes

to the area of a dime held at arm's length. Many people are surprised to learn that only such a small fraction of the field of view of each eye is seen clearly at any instant, but you can prove it to yourself by closing one eye and fixing the other on some small object, say a light switch across the room. You will notice that even objects quite close to the one you are staring at seem indistinct and lacking in detail; remember that to perceive this effect you have to concentrate on keeping your eye fixed, since it naturally tends to wander around.

About 5 mm from the axis of the eye on the nasal side, the optic nerve leaves the retina, which causes an interruption in the surface that produces a blind spot about 1 mm in diameter. This arrangement means that the right eye has a blind spot on the retina to the left of the axis, whereas the left eye has one to the right. However, it is very important to realize that when projected out into the field of view, the situation is reversed: the right eye has a blind spot in the field of view to the right of the axis and the left eye has one to the left. The reason for the reversal is not difficult to understand if you remember that a lens forms a real image that is inverted *and* reversed right to left. Therefore, objects to the right of the eye's axis in the field of view are imaged to the left on the retina and thereby fall on the blind spot. Figure 8–5 illustrates this effect.

Again, people are often surprised to learn that, in addition to not seeing clearly over most of the field of view at any instant, they do not see at all over a part of the field of view! Usually the blind spot is not noticed because that portion of the field of view of the right eye is seen by the left eye and vice versa. In addition, the eyes normally continually scan the field of view (more about eye movements later), so no

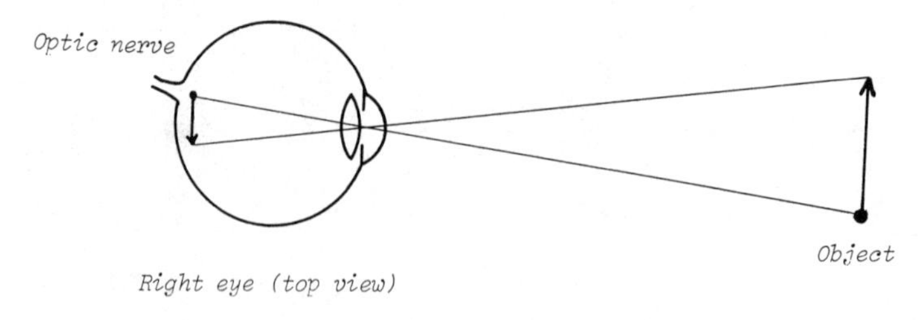

Figure 8–5 Imaging on the blind spot

Figure 8–6 Blind spot demonstration

point is unseen for very long even by a single eye. The presence of the blind spot can be simply demonstrated with the aid of Fig. 8–6. You must close the left eye and fix the stare of your right eye on the cross (black dot to the right). Now move the book from arm's length toward your eye (keep it fixed on the cross). At a distance of about 18 cm the dot should disappear! Note that you do not perceive a dark spot at the blind spot, just as you do not perceive darkness at the back of your head; you simply do not see anything.

8.4.3 Rods and Cones

So far we have said that the retina is the light-sensitive layer in the eye without giving any more detail about how it works. Much more is known. On a microscopic level we know that the retina is made up of two types of light-sensitive cells, called **rods** and **cones,** along with their nerve connections. Figure 8–7 shows simplified drawings of a rod and a cone and clarifies why they were so named. Each retina has about 120 million rods and 6 to 7 million cones packed into it.[8, 9] You can see that they must be

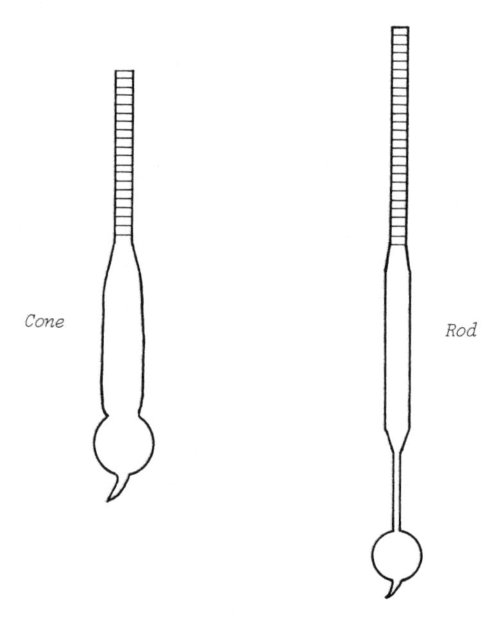

Figure 8–7 Human rod and cone

[8] Leo Levi, *Applied Optics: A Guide to Optical System Design,* vol. 2, (New York: John Wiley & Sons, Inc., 1980), p. 349.

[9] Mathew Alpern, "The Eyes and Vision," in *Handbook of Optics,* ed. Walter G. Driscoll (New York: McGraw-Hill Book Company, 1978), pp. 12–21.

very tiny, about 0.05 mm (or 50 μm) long and only 1 to 3 μm in diameter. This small diameter is necessary for high-resolution (fine-detail) vision because each cell can only signal how much light is falling on its entire cross-sectional area; no finer variations are detected. However, this size is well matched to the imaging ability of the eye, since the real image formed will not have any clear details smaller than about 2 μm anyway. The cones are concentrated in the fovea, which contains no rods, whereas the outer portion of the retina contains mostly rods. Of course the blind spot contains neither rods nor cones.

The two types of receptor cells have different functions. The cones are responsible for high-light-level, high-resolution, color vision, while the rods handle low-light-level (or night) vision. These two kinds of vision are named **photopic** and **scotopic,** respectively. The change between photopic and scotopic vision is an automatic one, triggered by the ambient light level and called **adaptation.** Full adaptation from photopic to scotopic, say in going from daylight into a darkroom, can take 20 to 30 min and result in an increase in average light sensitivity by a factor of more than 10,000.[10] Adaptation from scotopic to photopic is accomplished much more rapidly, in a few tenths of a second.

How can such a large range of sensitivity be accomplished by the eye? We mentioned earlier that the pupil size adjusts to the light level. But this adjustment cannot be the full story because pupil diameter only changes at most by a factor of 4, giving a 16-fold increase in area and collected light; we are trying to explain much larger increases in sensitivity. Also we know that the pupils of both eyes change together, but each eye can be adapted to quite different light levels. You can prove this to yourself. Start with both eyes dark-adapted. Close and cover one eye when you turn on a light at night, while keeping the other open. When you turn out the light after a few seconds and try the vision in each separately, you will find that the one left open is now almost blind, but the one left closed still has reasonable night vision. So pupil dilation is just a small part of dark adaptation or scotopic vision. We must seek the answer in a closer examination of what happens inside the rods and cones themselves.

Although no one is certain of the process, it is thought the light that enters one of the rods or cones first causes a chemical change in a substance inside called a **visual pigment.** The rods have one kind of pigment, while each cone has one of three different kinds to distinguish colors (more about color vision later). The first step is a bleaching of the visual pigment from a darker state to a clear state by the light. Somehow, not yet understood, this change gets transformed into an output in the form of a nerve impulse from the other end of the cell. For understanding vision it is important to realize that it is the change in the visual pigment that produces the output of a rod or cone; once the pigment is bleached out fully, it must be regenerated in an unbleached state before the cell can respond again. Now apparently the visual pigment in the rods is much more sensitive to light than those in the cones. Accordingly the rods "saturate" (all visual pigment bleached) at much lower light levels also. Therefore adaptation from scotopic to photopic vision consists of the rapid, com-

[10] Ralph M. Evans, *The Perception of Color,* (New York: John Wiley & Sons, Inc., 1974), p. 19.

plete bleaching out of the rod pigment, so that the rods no longer respond, whereas the adaptation from photopic to scotopic involves the slower regeneration of that pigment at low light levels. Scotopic vision is usually considered to extend from light levels of starlight (clear, moonless night) to a level around that of quarter-moon light. Photopic vision (rods completely saturated) extends from twilight to full sunlight. In the lighting range from twilight down to somewhat above quarter-moon levels, an intermediate situation exists in which there is not enough light to completely saturate the rods but there is enough to activate the cones weakly. Because both systems are active in this illumination domain, this type of vision is often called **mesopic.** Mesopic vision seems to give rise to an eerie, somewhat disturbing sensation in people. When there is a full moon mesopic vision persists all night outdoors, and perhaps its unsettling psychological effects have something to do with the "lunacy" attributed to the effect of the moon in folk wisdom.

8.4.4 Spectral Response

In view of the above description, which should be considered tentative and subject to change by new experimental evidence, it should not be surprising that the two types of vision exhibit a different spectral (or wavelength) response. These two different spectral responses are shown in the graphs of **relative luminosity** or **spectral luminous efficiency** in Fig. 8-8. These curves, which show the relative ability of different wavelengths of light to produce the brightness sensation, represent the culmination of years of psychological experiments with many observers. The photopic curve was adopted as standard in 1924 and the scotopic in 1931 by the Commission Internationale de l'Éclairage (CIE), an international standards commission for optics. The

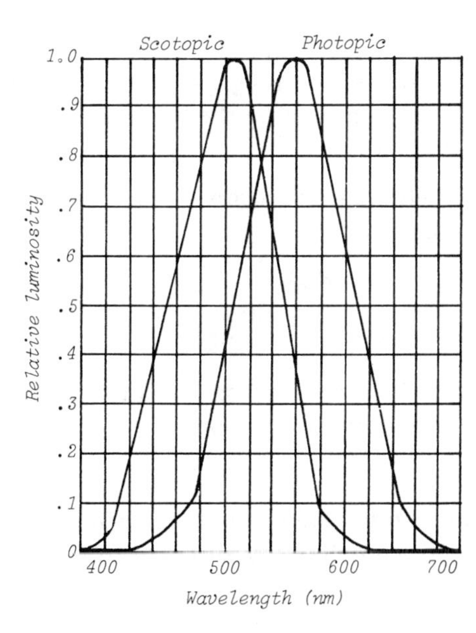

Figure 8-8 Relative luminosity: photopic and scotopic

relative nature of the curves is indicated by the peak value of 1 for both of them: they show the brightness-evoking capability of each wavelength *relative* to that of the most effective wavelength at the peak of each curve. This most effective, or **peak,** wavelength is 507 nm for scotopic vision and 555 nm for photopic vision. For any wavelength light, multiplying its spectral luminous efficiency value (K_λ) by its power gives a measure of its relative brightness.

Example:

A cadmium laser has an output of 40 milliwatts (mW) at a wavelength of 442 nm, and a helium-neon laser has a 2-mW output at 633 nm. Which appears brighter to a light-adapted observer who views both reflected from a sheet of white paper?

We go to the photopic relative luminosity curve of Fig. 8–8 to read $K_{442} \cong 0.03$ and $K_{633} \cong 0.25$.

Then relative brightness for the Cd laser is

$$0.03 \times 40 = 1.2$$

for the helium-neon laser $0.25 \times 2 = 0.5$

Since $1.2 > 0.5$, the cadmium laser appears brighter. Furthermore, we know it should appear $1.2/0.5 = 2.4$ times as bright.

Example:

How many times more effective is light of wavelength 555 nm than light of wavelength 510 nm in producing the brightness sensation in an observer using photopic vision?

This question asks precisely what the relative luminosity curve tells us. Again reading off the graph:

$$K_{555} = 1$$
$$K_{510} = 0.5$$
$$K_{555}/K_{510} = 2$$

So the 555 nm light is exactly twice as effective (watt for watt) as the 510 nm light in producing the brightness sensation.

The different curves for scotopic and photopic vision are indicative of the fact that the visual pigment of the rods is a substance different from that of the cones; actually the photopic curve probably represents the composite effect of three visual pigments.

A number of other interesting facts should be noted concerning the two types of vision. First, since the fovea contains only cones, it is night-blind. To get the best view of something under low-light-level conditions, you should use averted vision, that is, look slightly away from the object of interest in order to move its image off the fovea. On the other hand, the rods are color-blind; one visual pigment provides no way to distinguish among wavelengths. Thus scotopic vision is monochromatic, like a black and white television, leading to the saying that "all cats are gray at night." However it is important to remember that the whole cone color-vision system is present and ready to operate under low light levels; the problem is just that there is not enough light to stimulate it. This fact is confirmed by the observation that a small, bright enough, colored light (say a lighted cigarette or the numerals on a light-emitting diode (LED) digital clock) is seen as colored even by the dark-adapted eye. Also it is impossible to see fine details under scotopic vision to matter how close the object is brought to the eye. Evidently the output of many rods is summed in the retina before being sent on in order to gather more light.

The shift in spectral response toward shorter wavelengths with the onset of scotopic vision is known as the **Purkinje effect** after its discoverer, Johannes Purkinje (1787–1869), a Czech scientist. It means that blues and greens will become brighter relative to reds and yellows as the light fades. For someone who likes to sit outside through a summer sunset into twilight, a geranium plant will make a good subject for an experiment to personally verify this effect. Under good light the red geranium blossoms are much brighter than the dark green leaves. As twilight deepens, all the colors fade toward gray but the red fades more rapidly, eventually reaching a darker shade than the leaves, provided there is no artificial illumination on the plant. As a matter of fact Purkinje himself first made this observation on blue and red flowers at dawn.[11] Along these lines, we should also note that since the two eyes may be differently adapted, the Purkinje effect can be even more dramatically demonstrated between the two eyes so adapted by alternately opening one and closing the other while viewing different colors in a darkened setting. Clever use of the Purkinje effect also allows one to use a light in the dark but still retain scotopic vision—the trick is that the light must be a deep red. A glance at Fig. 8–8 shows that if almost all the power of the light is concentrated above about 600 nm in wavelength, then it can stimulate the cones without affecting the rods to any great extent. Thus astronomers may use a red-filtered flashlight to read their telescope settings at night (cone vision) without ruining their dark adaptation for seeing faint objects through the telescope (rod vision). The evolutionary reason for the Purkinje effect seems to lie in the fact that the spectrum of natural light shifts at twilight toward the blue of the sky, since the yellow-white of direct sunlight is absent after sundown.

[11] *Light and Color,* (Cleveland, Ohio: General Electric Lighting Business Group, 1978), p. 11.

8.4.5 Efficiency of Light Use

It is interesting but rather strange that the cells seem to be in the eye backward. The light-sensitive portions, the slender tops in Fig. 8–7, face away from the light toward the choroid at the back of the eyeball. Therefore the light must pass through several layers of nerves and the cell bodies themselves before it can be sensed. Indeed, there would be no need for an interruption in the retina and a corresponding blind spot if it were not for this awkward arrangement with nerve connections on the front of the retina instead of behind. Several explanations have been offered. Some have said that it is the natural result of the embryological development of the retina as an outward extension of the brain,[12] although why the backward placement of the cells is a natural result has never been made clear. It is true that many scientists consider the retina as an extended portion of the brain, and incidentally, the only such portion readily accessible to visual examination. A more recent theory takes note of the fact that the light-sensitive tips of the rods and cones are formed by stacks of disclike plates. These plates are formed at the base, where the tip joins the fatter cell body, and they are shed as cellular debris at the end of the tip; the light-sensitive column is continually growing outward from the cell body. This theory[13] suggests that the cells are in backward so that these dead plates can be swept away by circulation behind the retina, thereby preventing a light-disrupting accumulation in the vitreous humor. Figure 8–9 shows a schematic diagram of the layers in the retina.

To follow the light to the bitter end, that which passes through the retina without being absorbed by it is absorbed in the dark choroid behind the retina and contributes no more to the process of vision. There are some animals that have a reflecting layer (composed of much the same sort of material as fish scales) immediately behind the retina to pass the unabsorbed light back through the rods and cones. Such animals have eyes that seem to glow in the dark when they look into the headlights of your car or a flashlight you are holding. The optical system of the eye which focuses incoming light on the retina also works in reverse to project any light

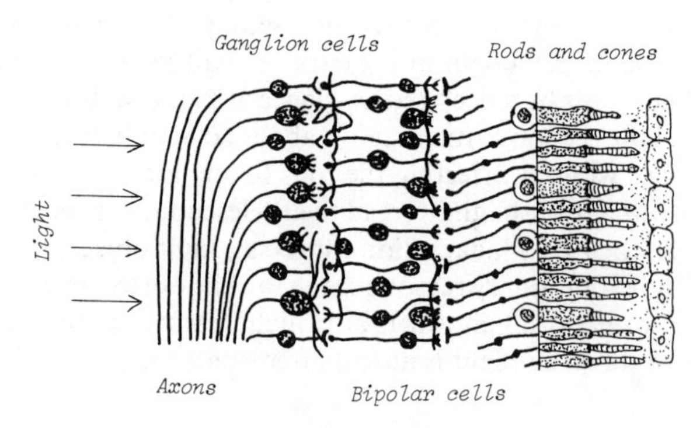

Figure 8–9 Retinal cross section magnified

[12] Gregory, *Eye and Brain,* p. 61.

[13] Levi, *Applied Optics,* p. 350.

reflected from the retina back in roughly the direction it came from; in other words, because of the reversibility of light rays the eye acts as a retroreflector insofar as it reflects at all. Under ordinary circumstances you cannot see into another person's (or animal's) eye; the pupil looks black because your head is blocking any light source that would cause retinal reflected light in your direction. In order to examine a patient's eye a doctor must use a device, the **ophthalmoscope,** that projects a small light beam along his or her line of sight by means of a partially silvered mirror. In the case of a dog's or cat's eyes shining in your headlights however, so much light is being reflected that you can see them against a dark background even though your line of sight is somewhat off the beam of light. The same cause gives rise to the notorious "pink eye" effect known well to photographers. When a flash color photograph is taken of a person looking at the camera, the recording eye is the camera. If the flash goes off close to the camera lens, often the subject's portrait will show eyes with distinctly pink pupils. What has happened in this case is that the subject's eyes were flooded with so much light that even the small percentage reflected from the retina is enough to expose the film, and furthermore since it is reflected in the general direction of the flash, enough enters the nearby camera lens. Simply put, the retina, pink from its blood supply, has been photographed. The solution is simply to move the flash away from the camera lens: the same amount of light will be reflected by the retina toward the flash, but now away from the camera.

Estimates are that roughly 50 percent of the light entering the human eye reaches the retina (the other half being reflected or absorbed by the structures in front), and of that half some 10 to 20 percent is absorbed in the retina, thereby contributing to vision. Although those percentages do not seem very high, the light that is absorbed in the retina is used with incredible efficiency. At the extreme lower threshold of scotopic vision and at the peak wavelength of 507 nm, practically every photon is recorded by the visual system.[14]

8.4.6 Temporal Response

We noted earlier that the outputs from many light-sensitive cells may be summed before the information is passed on. This process amounts to integration of the light over a spatial area of the retina, leading to increased sensitivity at the expense of reduced spatial resolution. We should now also note that the light is integrated over a time period known as the **integration time, action time,** or **storage time** of the eye. This integration time is estimated to be from 0.1 to 0.2 s.[15, 16] Since the light from the rods and cones is effectively summed and stored over this time interval, variations in the scene that occur in a much shorter time will not be perceived; 0.2 s can be roughly considered as the response time of the eye. This time seems to be matched closely to the response time of the whole human system of nerves and muscles. Even if we

[14] Rose, *Vision*, p. 41.

[15] Ibid., p. 50.

[16] *The Science of Color*, p. 109.

could see changes more rapid than 0.2 s, our bodies could not react in time to take advantage. Here the visual system gains increased sensitivity to light at the expense of reduced resolution in time.

Related to this response time of the eye is the **critical fusion frequency** or **critical flicker frequency** (CFF). When the eye views a periodically changing light source, the source can be seen as flickering if it is not changing too rapidly. The minimum frequency of the source which produces no flicker sensation is the CFF. The value of the CFF ranges from 5 to 55 Hz, depending on the brightness and size of the source. The CFF of the human eye is an important factor in determining the frame time of motion pictures and television, each of which consists of a rapid succession of still pictures or frames. Motion pictures only change frames 24 times per second, but a rapid shutter interrupts each frame several times, so the eyes see more than 60 frames per second and consequently no flicker. In television only enough information for 30 frames per second is transmitted. The flicker problem is solved by painting out every other line in a frame in one-sixtieth of a second; in the next sixtieth of a second the alternate lines are painted out. Then, as far as the eye is concerned the scene is changing 60 times per second and no flicker is seen. We might say 30 frames per second are shown by showing 60 half-frames per second. Even ordinary household lighting has a variation in it due to the 60 Hz alternating current that drives it, but this frequency also is above the human CFF, so that no flicker is seen.

8.4.7 Information Processing

Once the rods and cones in the retina have done their job, the nature of our narrative changes. Now the process of vision involves electrical signals instead of light, and the setting switches to the nerve layers in the retina, as shown in Fig. 8–9. In general the rods and cones connect to bipolar cells, bipolar cells connect to retinal ganglion cells, and the ganglion cells connect to the brain. The retinal ganglion cells are much like ganglion cells elsewhere throughout the nervous system, having one long branch called the **axon.** It is the collected axons of the retinal ganglion cells that pass out of the eye as the **optic nerve.** Somehow visual information must be passed from the receptor cells to the bipolar cells to the ganglion cells to the brain. But this transmission certainly does not occur in a linear, one-to-one fashion. As indicated in Fig. 8–9, each receptor cell is connected to several bipolar cells and vice versa. Generally the same holds true for the connections between bipolar and ganglion cells. Only foveal cones have corresponding ganglion axons in the optic nerve, which may be considered "private lines" to the brain.[17, 18] Even these cones however also make additional diffuse connections. For the retina as a whole, the 127 million receptor cells send all information to the brain over less than 1 million fibers in the optic nerve.[19]

[17] Ibid., p. 92.

[18] W. D. Wright, *The Measurement of Colour,* 4th ed., (New York: Van Nostrand Reinhold Company, 1969), p. 37.

[19] Levi, *Applied Optics,* p. 349.

Therefore the network of nerves in the retina must be more than just lines for the receptor cells; enormous amounts of information processing must take place right in the retina, justifying the picture of it as an extension of the brain. In addition, under very low light levels the receptor cell and nerve network in the retina act to give enormous amplification, by a factor estimated to be greater than 1 million, for the energy in a nerve pulse is that much greater than the minimum detectable light energy.[20]

8.5 PATHWAYS TO THE BRAIN

After leaving the eyes, somewhere behind the midpoint of the two eyes both optic nerves divide into two bundles, so that portions of the optic nerves from both eyes travel to the left half of the brain and other portions from both eyes travel to the right half of the brain. This dividing and crossover point, known as the **optic chiasma,** is shown schematically in Fig. 8–10. The division is such that the fibers from the left half of each retina go to the left half of the brain, while those from the right half of the retina go to the right half of the brain. Now those readers who have some knowledge of the central nervous system may remember that the left half of the brain generally controls the *right* half of the body and the right half of the brain the left half of the body. The eye-brain pathways seem to be an exception, but perhaps not so much of an exception as at first appearance, for we must remember (as we did in considering the blind spot) that the retinal image is reversed right-to-left with respect to the scene in the field of view. Therefore the left half of the brain receives information from the right half of the field of view of both eyes, and the right half of the brain processes the left half of the field of view of both eyes.

As seen in Fig. 8–10, the axons from the retinal ganglion cells end at midbrain, where they make connections with new ganglions. At this point additional information processing takes place—again it is not just one-to-one connections. The axons of these new ganglion cells extend to the vision centers at the back of the brain, the

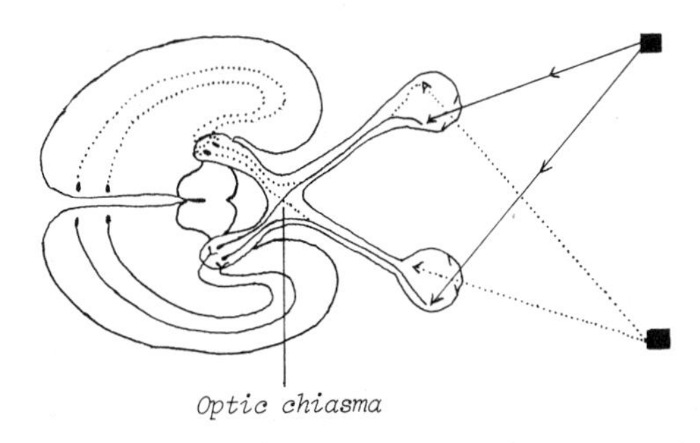

Figure 8–10 Neural pathways from eye to brain

[20]Rose, *Vision,* pp. 44–51.

visual cortex. Here is where the wonderful and still mysterious process takes place that produces the conscious "picture" we see.

We have traveled rather far afield from the eye, tracing the pathways of light and the electrical signals it produces, but the essence of vision seems still to have eluded us. That secret lies locked in the incredibly complex details of the visual cortex and its extension, the retina. The surface knowledge that we do have is a story for the next chapter.

QUESTIONS

1. Which element of the eye is the main refractive element?
2. What is the range of pupil size in the human eye?
3. Use Fig. 8–2 to find what pupil diameter is expected at twilight.
4. What is meant by the *consensual pupillary reflex?*
5. What function does the crystalline lens of the eye have?
6. For what object distance is the eye focused when relaxed?
7. What is the near point? What is its standard value?
8. What is presbyopia?
9. Explain what a cataract is and how it is treated.
10. Roughly what is the value of the effective focal length of the eye?
11. What is the light-sensitive layer in the eye called?
12. What is myopia and how is it corrected?
13. What is hyperopia and how is it corrected?
14. What is the area of the retina called that has the greatest visual acuity?
15. Where is the blind spot in the right eye's field of view?
16. Where is the blind spot on the retina of the left eye?
17. Why does the eye have a blind spot?
18. What are rods and cones? Which are more numerous?
19. Which type of receptor cell does the fovea contain?
20. Which is more sensitive to light, photopic or scotopic vision? How much more?
21. What is meant by mesopic vision?
22. Under photopic vision, how many times more effective is light of 520 nm than light of 620 nm wavelength in producing the brightness sensation?
23. Which is brighter, 5 mW of sodium light of wavelength 590 nm or 10 mW of helium-neon laser light of wavelength 633 nm (photopic vision)?
24. Why is averted vision necessary to see an object well on a dark night?
25. What is the Purkinje effect?

26. What is "pink eye" in color photography and how can it be avoided?
27. What is the action time of the eye and what is its value?
28. What is the critical fusion frequency and what is its value?
29. What is the optic chiasma?
30. The right half of the visual cortex is injured. What part of the right eye's field of view is likely to be affected? The left eye's?

Chapter 9

Seeing

In this chapter we consider some of the general properties of the visual system that can be demonstrated by psychological or psychophysical experiments. Presumably scientists will someday be able to explain these properties and others yet undiscovered by analysis of the detailed working of the brain-retina. Some studies have already been done with animals which relate elements in the field of view to signal patterns at various places in the neural pathway to the visual cortex.[1] However, such studies are far from providing a complete explanation of vision, and even to consider them in detail would take us too far from the purpose of this text.

9.1 EYE MOVEMENTS

We see with the visual cortex in the sense that here is where the conscious picture is formed. What we see is not the retinal image; rather we see with the aid of the retinal image. This basic understanding comes from our knowledge of the foveal eye and experiments on eye movements. We have already said that only a very small fraction (about one-thousandth) of the field of view falls on the fovea and is consequently seen in sharp detail at any instant.[2] Also we have briefly noted the eye's tendency to scan the field of view. One might think that the eye scans around until the whole field of view has passed over the fovea and been registered. But that is not what happens at

[1] David H. Hubel, "The Visual Cortex of the Brain," *Scientific American,* November 1963.

[2] E. Lleywellyn Thomas, "Movements of the Eye," *Scientific American,* August 1968.

all. The movement of a subject's eye can be recorded by reflecting a small light beam off the cornea to a filming camera. Then these movements can be correlated with the scene viewed. It is found that when generally scanning or searching a scene, the eye moves between momentary fixations in rapid movements called **saccades.** The saccades take only a few milliseconds, while the fixations last a few tenths of a second, depending upon the interest of the detail fixed. It is not surprising that the fixation time is comparable with the integration time of the eye (0.1 to 0.2s), since a much shorter time would not take full advantage of the eye's light-gathering ability and a much longer time would bring diminishing returns. The saccades and fixations are subconscious, showing a rather random pattern, but they can also be consciously overridden to follow a moving object smoothly or to stay fixed on a stationary object. However, even when the eye is consciously fixed for a long period, it undergoes small subconscious movements of drift, rapid correction, and a super-imposed high-frequency (30 to 80 Hz) tremor.[3]

9.1.1 Stabilized Retinal Images

Apparently constant eye movement and corresponding movement of the real image across the retina are necessary for correctly functioning vision. Ingenious experiments have been conducted with a microprojector on a contact lens fitted right on the eyeball.[4] An image is projected into the eye which moves with the eye, and therefore the image is stabilized on the retina. When an image is stabilized in this manner, the subjects report that the scene viewed begins to fade out after a few seconds; parts of the picture disappear one by one until only a featureless gray field remains. Continued viewing brings back parts of the scene, which now keep reappearing and disappearing. One interesting feature of this phenomenon is that the disappearance or reappearance takes place in meaningful chunks. Thus the word BEER might become PEER or PEEP or BEE or BE, and pictures lose or gain coherent parts. Here again change in stimulus seems to be a vital feature in making the visual system work, although more than bleaching of the visual pigments in certain receptor cells must be involved.

9.1.2 Fixations

To return to the question of free scanning of a visual scene, the experiments have shown that only a small part of any scene has been seen clearly, even when subjects are instructed to study it closely and given all the time they want. A series of fixations and saccades are made and then the subject reports having studied the whole scene; only the record of fixations, along with our knowledge of the area of greatest visual acuity, clearly reveals that most of the scene was recorded by the subject's retina in a

[3] Mathew Alpern, "The Eyes and Vision," in *Handbook of Optics,* ed. Walter G. Driscoll (New York: McGraw-Hill Book Company, 1978), pp. 12–18.

[4] Roy M. Pritchard, "Stabilized Images on the Retina," *Scientific American,* June 1961.

decidedly indistinct manner. Within the scene itself, lines or edges seem to attract a disproportionate share of fixations. This result is in accord with the neural signal studies on animals mentioned earlier, and we shall see that other visual theories stress the importance of edges. In fact, it might be discouraging to artists to learn that some studies of subjects viewing artwork found that the picture frame or boundary drew more fixations than points within. Anything moving or flashing, particularly in peripheral view, draws many fixations. Even more interesting is the fact that people often deny having seen some detail of a scene with emotional content despite instrumentation indicating that they indeed did look at it. Are the subjects lying to the experimenter or perhaps to themselves, willing themselves to forget something distasteful they saw? Probably not, for other experiments have shown that we do not see in a conscious sense all the details we look at (image on the fovea), even in scenes devoid of any obvious emotional content. Subjects who are asked to search, under a time limit, for one number in a large table of numbers sometimes report not finding it even though the camera recorded a fixation upon the very number, maybe even more than once! It is hard to imagine a more unemotional scene, and the subject has nothing to gain by deceiving the experimenter or him- or herself.

By now the reader may be justifiably confused. If the retina cannot send detailed information on most of the scene and our conscious picture may be missing detailed information the retina did pick up, just how and what do we see? An Arabian proverb[5] seems to sum up the situation nicely, "The eye is blind to what the mind does not see." To a great extent we see what we expect or want to see. It seems as if the visual cortex constructs a conscious picture from the information in a few fixations, and later fixations are used to refine the picture. In other words, our brains build a model of the world from limited visual information; the information from successive fixations provides clues that aid in the model building. The brain has a marvelous filling-in capacity to make up for the lack of detailed information on most of the scene, as we have already seen in the blind spot demonstration. It must be remembered that the model construction process is subconscious; all we "see" is the finished result in the visual cortex, not the snippets on the subconscious cutting-room floor.

9.2 OPTICAL ILLUSIONS

Optical illusions are often studied to throw light on the process from a different angle. In these cases the visual system is somehow being misled into seeing incorrectly. Since almost everyone sees them in the same incorrect way, maybe they can tell us something general about human vision. One of the best-known and most striking of optical illusions is the Müller-Lyer illusion, shown in Fig. 9–1.

Everyone sees the vertical line on the left as being longer, but really both vertical

[5] Ralph M. Evans, *The Perception of Color* (New York: McGraw-Hill Book Company, 1974), p. 20.

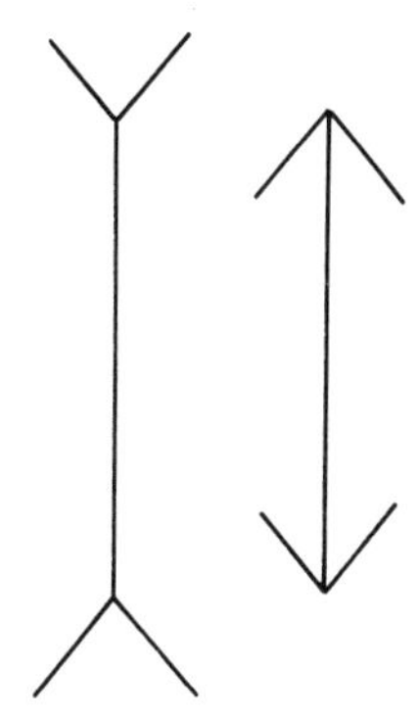

Figure 9–1 Müller-Lyer illusion

lines are the same length. Because the illusion is so striking, the reader is urged to check the two lengths with a ruler. The remarkable fact about this illusion and others is that even when you know the correct result, that the two lines are the same length, you still see it incorrectly, with the left line longer. Another example, perhaps less striking, is the Ponzo illusion, shown in Fig. 9–2. In this example the upper horizontal line looks longer, even though again both lines are the same length. One more example will suffice to begin our study, and that is the Poggendorf illusion of Fig. 9–3. Here the illusion lies in the fact that the diagonal line looks like two segments of different lines, with the line to the lower left appearing to be too far displaced to the left to connect with the one to the upper right. That they do connect smoothly when extended can be proved by laying down a straightedge along either.

The theory[6] that seems to best explain these illusions takes cognizance of the fact that the visual system is really designed to see in three dimensions rather than the two dimensions of a flat piece of paper. Two features of our three-dimensional visual world are perspective and size constancy. The latter means that as an object moves

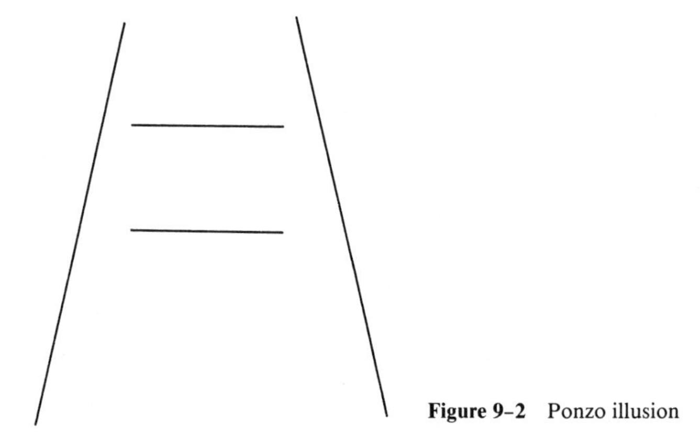

Figure 9–2 Ponzo illusion

[6] R. L. Gregory, *Eye and Brain: The Psychology of Seeing,* 3rd ed. (New York: McGraw-Hill Book Company, 1978), p. 150.

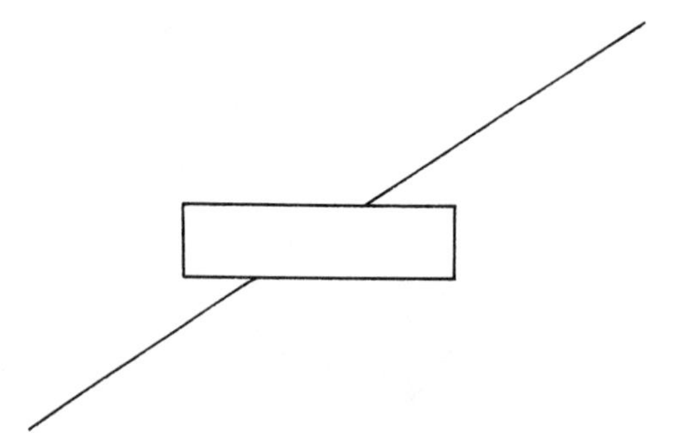

Figure 9–3 Poggendorf illusion

nearer or farther away from us, we do not see it as getting larger or smaller but rather as keeping the same size while changing its distance. However, as far as the retinal image is concerned, the two cases are the same: an object the same size as a nearby one, but twice as far away, produces a retinal image one half as large, just as does a half-size object alongside the nearby one. Again the brain is clearly using the retinal image size as only one clue in constructing the visual world. A distant object that we recognize is automatically scaled up to near the size we know it to be. Try using your thumb and forefinger at arm's length to "measure" a distant person or automobile and then bring your "measured" length around until it superimposes on a corresponding nearby object; you will be astounded at the size disparity, which will be the same as that of the retinal images. This size constancy scaling effect, noted as long ago as 1637 by René Descartes (see Chapter 1), seems to be acting along with the similarity of lines in the illusions to perspective lines to mislead the visual system.

In the Müller-Lyer illusion the figure on the left looks like a perspective drawing of an inner corner, as of a room the observer is in, while that on the right looks like an outer corner. Accordingly the vertical line on the left, receding into the paper, would be farther away and the one on the right would be nearer. But since they both give the same size retinal image, the left line should be longer in the real world and so the brain scales it up. In other words, the brain is reading some perspective clues and then scaling the size of the objects accordingly. Of course you do not have to think all this out because it is done automatically and subconsciously as part of the model building process. In a similar manner, the slanted lines in the Ponzo illusion suggest horizontal perspective lines receding into the distance. Such an interpretation puts the upper horizontal bar farther away, so size constancy scaling by the brain makes it appear larger. The Poggendorf illusion makes sense if the slanted lines are horizontal in three-dimensional space and the rectangle is the vertical riser of a step.[7] Then the lower line is far displaced to the left to be in line with the upper one; the "correct" perspective drawing is shown in Fig. 9–4.

Few scientists would be rash enough to claim that the theory outlined above gives a complete explanation of all optical illusions. However it does seem to make a good beginning toward explaining many of the more prominent ones and it does fit

[7] Barbara Gillam, "Geometrical Illusions," *Scientific American,* January 1980.

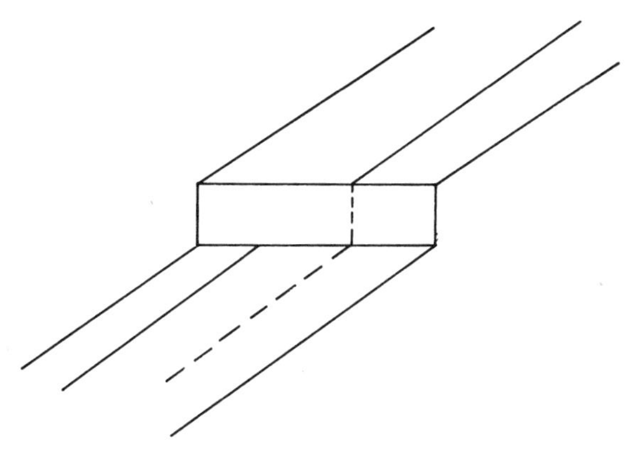

Figure 9–4 Perspective interpretation of Poggendorf illusion

nicely with the other evidence for the brain as a visual model builder. Furthermore there is additional support from studies of perhaps the most famous (and one of the earliest discovered) optical illusions of all, the Necker cube, shown in Fig. 9–5.

In this illusion the upper and lower square seem to alternate as the front and back face of a cube in three dimensions; the visual system cannot settle on either alternative because of the ambiguity of the figure. However it is obvious in this case that the visual system is trying to build a three-dimensional model from a two-dimensional pattern of lines, for otherwise there would be no illusion at all; if the figure interpreted as a two-dimensional pattern of lines, there is no possible ambiguity. It is interesting to notice that as the figure becomes more symmetrical, it is easier to see it as a two-dimensional pattern. This effect is illustrated by Fig. 9–6.

The Necker cube illusion in fluorescent, glow-in-the-dark paint has been shown to subjects in a darkened room in order to remove any perception of the two-dimensional surface on which it is drawn. Under these circumstances, the perception of a three-dimensional figure is even more pronounced, as is the illusion itself of alternating forward faces. But what is most important with regard to our theory of optical illusions is that the square which is temporarily seen as the back face is simultaneously seen as larger also.[8] Again, size-constancy scaling seems to be acting to "blow up" any object perceived as farther away but with the same retinal image size as an apparently nearer object.

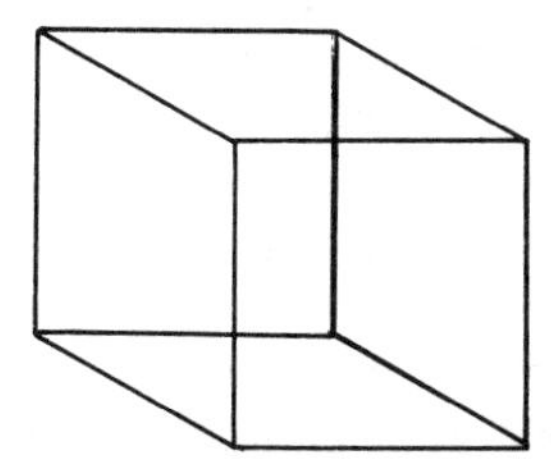

Figure 9–5 Necker cube

[8] Richard L. Gregory, "Visual Illusions," *Scientific American,* November 1968.

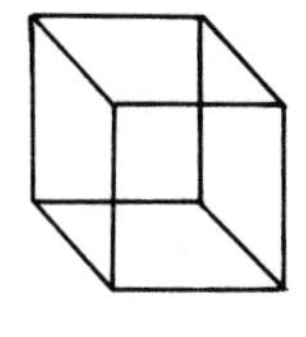

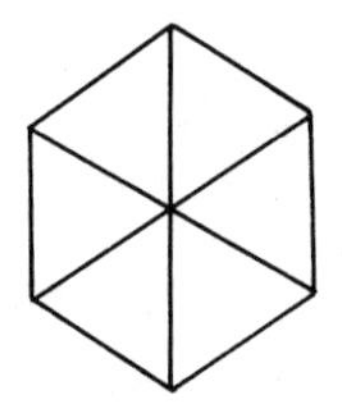

Figure 9–6 Necker cube from different perspectives

9.3 SEEING IN THREE DIMENSIONS

Considered superficially, optical illusions might seem to point out shortcomings in the human visual system. But having studied the source of the misperceptions we can conclude that the visual system is designed to see real, three-dimensional objects in a real three-dimensional world, which it normally does very well. Because it does this primary task very well, it can be fooled and indeed has been intentionally fooled for centuries by artists using perspective and other tricks, but that should not detract from our appreciation of its general effectiveness.

9.3.1 Monocular Depth Perception

In discussing optical illusions we have been led naturally to consider the problem of seeing in three dimensions, that is, depth perception. Only every consideration so far applies to a single eye working alone. We now need to look at depth perception in general and in particular at how having two eyes contributes. After all the retinal images themselves are two-dimensional, giving immediate information on up-down and right-left. How do we get our perception of near-far? Actually we have already discussed some of the ways that even a one-eyed person can get such information. Size constancy can work to give a rough perception of distance if a distant object is a familiar one; a small retinal image implies a great distance. This relation works both ways, since, as we have already seen, other distance cues may cause us to scale up or down an object from what its retinal image indicates. To judge distance from the retinal image size requires some familiarity with the actual size of the objects; thus distant mountains often seem closer than they really are because we are not used to looking at such immense objects and so interpret the relatively large retinal image as indicating nearness. We have also mentioned **perspective lines,** that is, lines which we have reason to believe are parallel (railroad tracks, edges of a highway, furrows or rows of crops in a field, etc.) but whose retinal images seem to converge toward the horizon. One reason that driving a large truck is easier than experience with driving an automobile would indicate is that the perspective view of the lane lines is much better from higher up.

There are several other single-eye, or **monocular,** depth cues that have not yet been mentioned. One that seems to combine some elements of the previous two is **texture gradient,** illustrated by Fig. 9–7. The texture involved might be quite random, like the grass in a meadow or the waves on a lake. Here there is no need for previous knowledge of the elements making up the texture, for those nearby serve as natural references for those more distant. In other words, the pattern taken as a whole has depth cues built in. Another obvious monocular depth cue is **overlap**—an object whose outline is partially obscured by a second object must surely be farther away. Even the amount of accommodation of a single eye provides a subconscious clue for distance, although this method is ineffective beyond about 6 m because the eye is relaxed and makes no accommodation for larger object distances. The monocular depth cue of aerial perspective is discussed in Chapter 10.

9.3.2 Binocular Depth Perception

More reliable depth perception, which works over longer distances, depends on the fact that human beings have two eyes separated by about 6.5 cm and looking in the same direction; this is **binocular depth perception.** It is important that the two eyes share a common field of view; many animals have eyes essentially on the sides of their head with little overlap in the fields of view, and so cannot have the binocular depth perception that humans have. Furthermore vision in one eye can be much worse than that in the other without seriously degrading this binocular ability.[9] Of course, blindness in one eye will eliminate binocular depth perception, forcing reliance on monocular depth cues. Perhaps it should be noted here that despite the obvious advantages of binocular vision, the performance of even a complicated task such as landing an airplane was not significantly impaired in experiments in which the vision of one eye was suddenly cut off.[10]

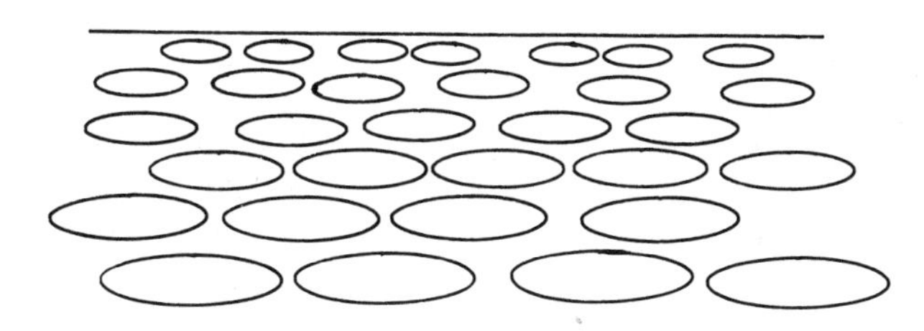

Figure 9–7 Texture gradient

[9] Leo Levi, *Applied Optics: A Guide to Optical System Design,* vol. 2 (New York: John Wiley & Sons, Inc., 1980), p. 421.

[10] D. Regan, K. Beverley, and M. Cynader, "The Visual Perception of Motion in Depth," *Scientific American,* July 1979.

There appear to be two components to binocular depth perception. One is **convergence,** the turning inward of the two eyes to look at an object (or bring it onto the fovea). An object closer to the observer requires more convergence, while one farther away requires less, as shown in Fig. 9–8. Somehow the convergence angle through which the eyes must turn is transmitted as a subconscious depth clue to the brain. This method works separately and independently on each object of attention in the field of view and is really identical in principle to the range finders used for focusing in some cameras.

The most remarkable and striking component to binocular depth perception depends on the fact that each eye sees the field of view from a slightly different angle and so sends a slightly different picture to the brain. This is known as **disparity depth perception** (because of the disparity in the two retinal images) or stereoscopic vision. The nature of this effect was first shown in 1838 by the English physicist Sir Charles Wheatstone (1802–1875) with the demonstration of his invention, the **stereoscope.** The stereoscope simply presents two different pictures of a scene from two different angles, one to each eye; the brain fuses the two images just as it would those from a three-dimensional scene to achieve a single startling picture that seems to jump off the two-dimensional plane. Figure 9–9 shows such a stereoscope. Wheatstone first used line drawings of objects from two different perspectives, since photography was just beginning at the time. Such a pair of pictures is called a stereo pair; one such pair of drawings is shown in Fig. 9–10 (note the perspective change). Within a few years he was using stereo pairs of daguerreotypes (an early form of photograph) in his stereoscope, which gave an even more realistic, rounded, space-filling picture. Figure 9–11 shows a stereo pair of photographs (again note the perspective change). Using such photographs the stereoscope became a popular parlor entertainment device in the latter half of the nineteenth century.

It is still not understood how the brain takes two different two-dimensional pictures and forms one three-dimensional picture, but apparently this process is the ma-

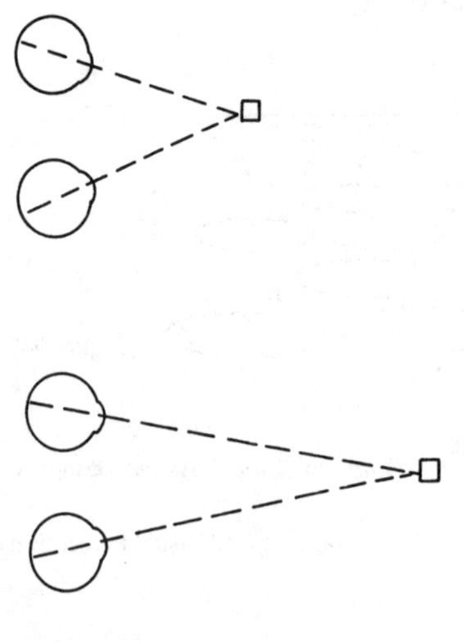

Figure 9–8 Convergence

Figure 9–9 Stereoscope (Photo courtesy of David Gustafson, Communication Resources, Inc.)

jor contribution to the fullness and depth we perceive in our visual picture of the real world. This fullness and depth still has the ability to surprise when we look through present-day children's stereo viewers (which use stereo pairs of photographic transparencies). Because our eyes are not very far apart compared with the scale of things we view in the real world, stereo vision has its limitations. The two retinal images from a very distant scene are not different enough to produce the stereo effect. Some authorities[11] say that about 100 m is the limiting effective distance, while others[9] say up to 1000 m (the latter refers to a stereo effect between an object at that

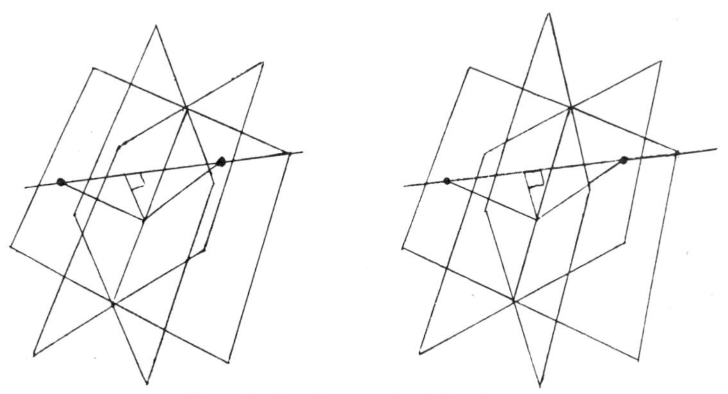

Figure 9–10 Stereo pair of drawings

[11] Gregory, *Eye and Brain,* p. 67.

Figure 9–11 Stereo pair of photographs (Photo courtesy of Arla Landrum.)

distance and a background much farther away). Even accepting the larger limit, we see that beyond 1 km we are forced back to using monocular cues.

9.3.2.1 Stereo Techniques. Over the years many ways have been found to utilize Wheatstone's basic trick in order to delight audiences with three-dimensional images. We can produce the stereo effect any time one eye is presented with an image and the other eye is presented with an image of the same scene from a different perspective. Making a stereo pair of photographs is not too difficult; stereo cameras are available that have two lenses separated by about the interocular distance and expose two frames at once (a stereo pair). Even with an ordinary camera you can take a stereo pair of photographs of a still life by taking one exposure and then moving the camera 6.5 cm (make sure to realign the same object in the center of the field of view) to take a second frame. Incidentally, as you might have guessed, moving the camera more than 6.5 cm will exaggerate the stereo effect. The more difficult part of the problem is finding a way to show one image of the stereo pair to the left eye and the other to the right eye. Some people, particularly after practice, can simply look at a stereo pair, such as Fig. 9–10 or 9–11, with the left image in front of the left eye and the right in front of the right eye and fuse them mentally into one stereo picture in the middle. You might try this yourself. However, most people need the help of a stereo viewer such as Wheatstone's stereoscope, which puts one picture at the focal point of one lens and the other at the focal point of a second lens. Then fusion into a stereo view comes readily when the two eyes are brought up to look through the lenses.

Stereo viewers such as the one described above have always had the drawback that only one person at a time can look. Other ways have been found to make the presentation to a larger audience. One method uses projected images and polarization of light. The left image might get projected onto a screen with vertically polarized light (by using a polarizing filter in the projector), while the right image is simultanteously projected onto the same screen with horizontally polarized light. Now each viewer must wear a set of spectacles consisting of a vertical polarization filter over the left eye and a horizontal one over the right eye. If a person views the screen without the polarizing spectacles, only a blurred two-dimensional image is seen, the superposition of the components of the stereo pair on top of one another. But with the spectacles on, the right image is blocked from the left eye (horizontal polarization blocked by vertical polarization filter) and the left image is similarly blocked from the right eye. Thus each eye only sees the correct one of the stereo pair, and a stereo image results. One American company that made stereo cameras also used to make and sell stereo slide projection equipment based on this polarization technique. This technique also was used for 3-D movies, which had quite a vogue back in the 1950s although the equipment had been invented more than a decade earlier. Like many other fads, 3-D movies died out rather quickly despite their startling realism; every such movie had an obligatory scene in which some object was hurled at the screen, causing the whole theater audience to instinctively duck. The reasons cited for the eventual failure of 3-D movies include the inconvenience of the spectacles (particularly for patrons who were already wearing eyeglasses) and the expense of conversion of the theater, which included an expensive and complex stereo projector and a special screen that would not affect the polarization of the light it reflected. One reason seldom mentioned, but important in this author's opinion, is that there was never a good movie made by this technique.

A similar technique for printed pictures uses color instead of polarization to separate the stereo pair components for the two eyes. In this case the elements of the pair are printed on top of one another with, for example, the left picture in red ink and the right picture in blue ink. Again, without a special pair of spectacles, the picture is rather an indistinct blur. The viewer must use spectacles with a blue left lens and a red right lens. Now when red ink on white paper is viewed through a red filter, it becomes invisible because the filter makes the paper appear just as red as the ink: the red filter removes any contrast between the ink and the paper. On the other hand the red filter also absorbs the blue light reflected by the blue ink, making it appear black. Similar reasoning shows that the blue filter makes the blue ink appear invisible and the red ink black. Therefore the blue left lens will reveal the left printed image (red ink) in black for the left eye, while making the right image (blue ink) invisible. The red right lens serves the same purpose for the right eye. Such a colored, printed stereo pair is called an **anaglyph.** The technique was even used to make 3-D comic books during the height of the 3-D movie craze and is still used today for some presentations of 3-D aerial photographic mapping. One obvious drawback is that only black and white final images are possible, whereas the polarization technique allows full color.

A more recent technique, which has been applied to 3-D postcards and pictures

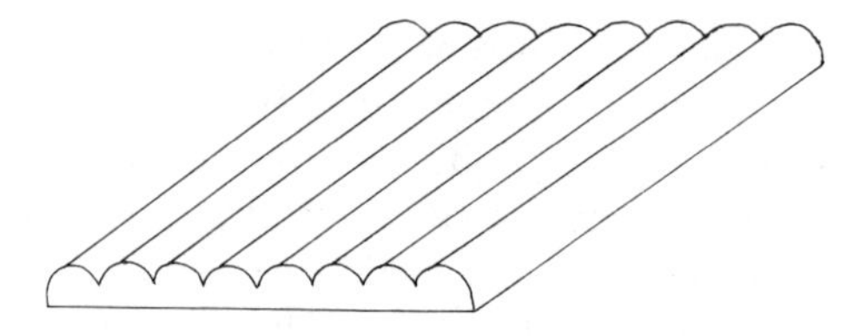

Figure 9–12 Lenticular screen magnified

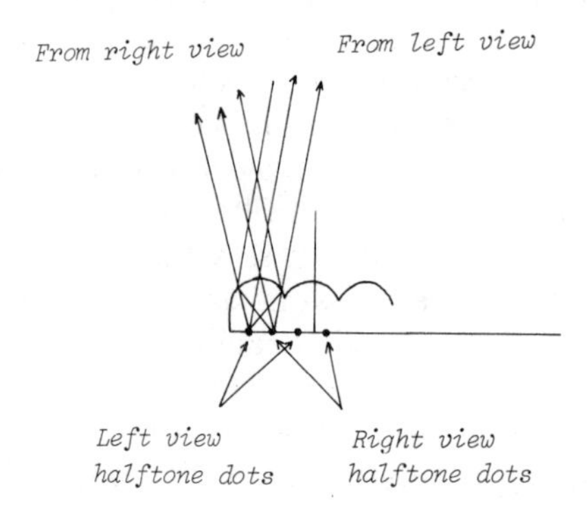

Figure 9–13 Interlaced stereo pair with lenticular screen

on children's book covers, uses a lenticular screen laid directly on top of the printed stereo pair. A **lenticular screen** is a piece of clear plastic that has been formed into the shape of a contiguous set of very small, parallel, cylindrical lenses. Figure 9–12 shows a magnified view. Such a screen characteristically has the appearance and feel of a flat piece of clear, striated plastic. If the two pictures of the stereo pair are printed in alternate, interlaced vertical lines and the cylindrical lenses of the lenticular screen, also vertical, are carefully aligned so a lens covers a pair of lines, then the two pictures will be projected out at different horizontal angles. Figure 9–13 illustrates this effect. Now each of the observer's eyes sees only one picture of the stereo pair because of the angular difference. This method is something like the old stereoscope but with the viewing lenses attached directly to the stereo pair. You can view such a 3-D picture with one eye while you move the picture horizontally, and you will see the two two-dimensional pictures of the stereo pair alternate as the projection of each is swept across your eye. With both eyes open you can see the same thing by turning the picture sideways (striations horizontal); now both eyes receive the same picture of the stereo pair and the stereo effect is lost.

9.4 COLOR VISION

Stereo vision adds a great deal to the beauty of our visual world compared with what it would be if flat. Probably even greater beauty is added by color vision, which we have barely touched on except to say that it is mediated by the cones in the eye. Color vision, like stereo vision, is not present in all animals. Apparently birds, fish, reptiles, and some insects do have it.[12] Strangely, however, there is no evidence for color vision in any mammals except the primates (monkeys, apes, and humans). Whatever else may make the bull charge, it is not the redness of the toreador's cape, and your dog or cat is oblivious to the colors of the pet toys you might buy.

What we have previously learned about color is that the different wavelengths of light give rise to the different colors of the spectrum (see Chapter 2). These colors, which can be produced by a single wavelength, are called **spectral colors.** Even before the wave theory of light had been established, Newton had shown that white light from the sun was composed of all visible wavelengths mixed together in roughly equal amounts; in this sense white light is a mixture of all the spectral colors. Please note that we are speaking of mixing *colored lights,* as in overlapping spotlights, not pigments. (For more on Newton's experiments and color mixing, see Chapter 11.)

The first solid clue as to how we see color though came from the discovery that all the spectral colors and white can be produced by the mixture of just red, green, and blue lights in various proportions. These three are therefore called **primary colors.** Figure 9–14 illustrates such mixing. In this figure, *R, G, B,* and *W* stand for red, green, blue, and white, respectively. *C* stands for cyan, a blue-green, which is the mixture of blue and green; *Y* represents yellow, the mixture of red and green, although it might be any shade from yellowish green to orange depending on the proportions of the two primaries. Moving through the various colors *R* to *Y* to *G* to *C* to *B,* we span all the spectral colors, and we have white at the center as a bonus. An additional bonus appears in the *R-B* overlap region, where *M* stands for magenta, a purple color. There is no single wavelength of light that will produce the sensation of purple, so magenta is a nonspectral color. Note that if we consider cyan, the overlap of blue and green, to be a separate colored light, then where it mixes with red, white is produced. Any two colors that will produce white when mixed are called **complementary.** So red and cyan are complementary, as are yellow and blue and green and magenta.

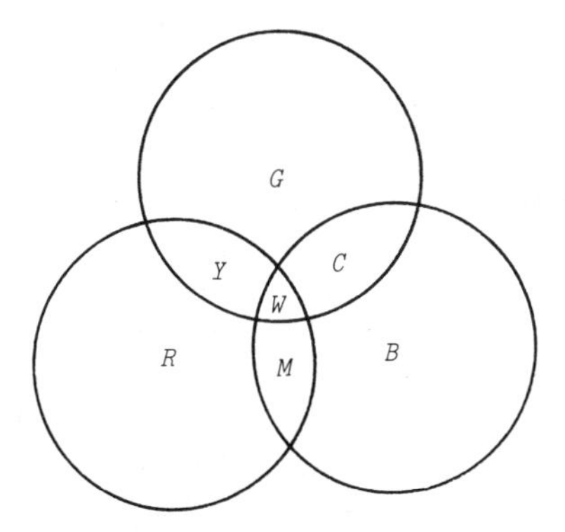

Figure 9–14 Mixing lights of the primary colors

[12] Ibid., p. 119.

9.4.1 Three-Pigment Theory

Thomas Young was the first to suggest a relation between these color mixing experiments and the properties of the human eye, thereby proposing a theory of color vision. He reasoned that it was improbable that the eye can contain a different receptor for each different perceptible spectral color (or wavelength), for that would require hundreds of different kinds; nature usually works more economically. Instead he proposed that the necessity for only three primaries in color mixing originated in the fact that there were only three kinds of receptors in the eye, one for each primary color. Other colors that we perceive would then result from a mixture of signals from these three kinds of receptors. Later the theory was put on a more quantitative basis by the German scientist Hermann von Helmholtz (1821–1894), who, like Young, was originally trained as a physician. Incidentally, Helmholtz was also the inventor of the ophthalmoscope, mentioned in Chapter 8. In more modern terms we speak of three different visual pigments with different spectral responses contained in different cones. Recent research has confirmed the presence of three such pigments.[13] This explanation of color vision is sometimes called the **trireceptor theory** or the **three-pigment theory**, or more commonly the **Young-Helmholtz theory.**

9.4.1.1 Color Blindness. The Young-Helmholtz theory provides a straightforward explanation of the various types of **color blindness.** Really the term **color blindness** is misleading, since a total lack of color sensations is extremely rare; rather we should probably refer to deficient or abnormal color vision. Strangely enough, no one knew that such a thing as color blindness even existed until 1794 when the great English chemist John Dalton (1766–1844) discovered that the color matches he and his brother would make showed differences from those everyone else made. Today we know that color blindness is a sex-linked, hereditary disorder that affects about 8 percent of the male and 0.4 percent of the female population. Color-blind individuals can be classed into three broad groups: anomalous trichromats, dichromats, and monochromats. These classifications are based on the number of primaries the subjects need in order to match all the spectral colors (to *their* satisfaction) in a light mixing experiment. Trichromats need three primaries, dichromats need two, and monchromats one. This definition implies that people with normal color vision are **trichromats.**

Anomalous trichromats are considered to have defective color vision because they mix the three primaries in abnormal proportions when color matching. This is the most common type of color blindness, affecting some 6 percent of the male population. The category breaks down further into three types: protanomalous, deuteranomalous, and tritanomalous. **Protanomalous** trichromats always use too much red (by normal standards) in mixing red and green lights to match a given yellow. On the other hand, **deuteranomalous** trichromats use too much green in the same experiment. The much rarer **tritanomalous** trichromats use too much blue in

[13] Edward F. MacNichol, "Three Pigment Color Vision," *Scientific American,* December 1964.

matching a cyan with a mixture of blue and green.[14] It is thought that the 1 percent of the male population who are protanomalous have a defective red-sensitive pigment in their cones, with a spectral response shifted toward that of the normal green-sensitive pigment. Because of this shift they are less sensitive than normal to red light and accordingly require more in their mixture to produce yellow. The 5 percent of the males who are deuteranomalous are thought to have a defective green-sensitive pigment, with a response shifted toward the red. Because both these trichromat types have too little separation between their red and green sensitivities, they have trouble discriminating among green, yellow, and red. A reasonable, but still speculative, assumption is that the tritanomalous trichromats have a blue-sensitive pigment with a spectral sensitivity shifted toward the green.

Dichromats divide into similar categories: protanopes, deuteranopes, and tritanopes. These people have a more profound color vision defect—they are apparently totally missing one visual pigment. **Protanopes,** about 1 percent of males, are missing the red-sensitive pigment; **deuteranopes,** also about 1 percent of males, are missing the green-sensitive; and tritanopes, far less than 1 percent of males are missing the blue-sensitive.[15] Protanopes and deuteranopes have far less red to green discrimination than the corresponding anomalous trichromats, having only yellow, blue, and gray sensations. The tritanope will have trouble discriminating in the blue-green region of the spectrum, with all wavelengths shorter than yellow looking blue-green and all longer appearing red (yellow itself should appear some shade of gray).[16, 17]

Monochromats are the only truly color-blind people (only 0.003 percent of males and 0.002 percent of females) in that they see the world in shades of gray only. They are either missing all three cone pigments (rod monochromats) or both the red- and green-sensitive pigments (cone monochromats). In the latter case we could say that the person has two different black-and-white visual systems, the rods and the blue-sensitive cones.

9.4.1.2 Afterimages. The phenomenon of colored **afterimages** is another element of color vision explainable by the three-pigment theory. When a bright colored shape is stared at for 30 s or so and the eye is then moved to stare at a neutral screen, an afterimage of the same shape appears but with a color complementary to that of the original. The reason is that the cones sensitive to the original color become fatigued (almost all the pigment or pigments for that color in that area of the retina is bleached out). When the eye is moved to a white screen, the cones sensitive to the original color cannot respond as strongly as the others, so white minus the original color shows up in that area of the retina. Along this line, in the stabilized image ex-

[14] *The Science of Color,* (Washington: Optical Society of America, 1963), p. 74.

[15] Alpern, "The Eyes and Vision," p. 12–25.

[16] *The Science of Color,* p. 136.

[17] W. D. Wright, *The Rays are not Coloured: Essays on the Science of Vision and Colour* (New York: American Elsevier Publishing Co., Inc., 1967), p. 93.

periments discussed earlier in this chapter color sensations were the first to disappear, which supports the presumption that a continuing stimulus can lower cone response.

9.4.2 Opponent Color Theory

The Young-Helmholtz theory explains the basic facts of color vision and seems to neatly account for the various categories of color blindness. Should we not then consider the question of color vision to be answered? Not quite, because there are some more subtle aspects of color vision that require deeper probing. In fact, the field of color vision has always been rife with controversy just because of some of these subtler aspects. The German psychologist Ewald Hering (1834–1918) used subjective properties of the unitary colors, red, yellow, green, and blue to construct his own theory of color vision. These unitary colors are the ones that *subjectively appear* unmixed, with their properties not derived from other colors, and so they have always played an important role in color theories. Hering noted that these unitary colors naturally group into pairs that never subjectively appear mixed with each other, red-green and yellow-blue. By this we mean that there is no color that appears reddish green or bluish yellow. He called the members of each pair **opponent colors,** and added black-white as a third opponent color pair. Hering theorized that there were three types of opponent color receptors in the eye, one type for each opponent color pair. Depending on the stimulating light falling on that portion of the retina, a receptor can indicate either member of its opponent pair *but not both.* Thus a red-green receptor can indicate the presence of either red light or green light, as a light switch can be either on or off, but not both at the same time. In this manner the opponent color theory explains subjective appearances by the nature of the receptors. Interestingly enough, Hering's theory appears to be quite valid, not at the receptors but further along in the visual system.[18] Experimental evidence indicates that color information is coded in the retinal ganglion cells into opponent colors; one ganglion may be able to transmit either red or green, while another transmits either yellow or blue. Such information organization, taking place right in the retina, still explains the subjective appearances, even though the Young-Helmholtz three-pigment theory holds at the receptor (cones) level.

9.4.3 Color Constancy and Simultaneous Contrast

Two other features of color vision that cause some difficulties for the Young-Helmholtz theory are color constancy and simultaneous contrast. **Color constancy** refers to the fact that we perceive object colors to be the same under quite different illumination. An apple appears about the same shade of red when viewed in direct sunlight (white), skylight alone (bluish white), or incandescent lighting (yellowish white). Yet we know that the light in the retinal image has a distribution of

[18] W. D. Wright, *The Measurement of Colour,* 4th ed. (New York: Van Nostrand Reinhold Company, 1969), p. 46.

wavelengths, or spectral composition, which depends on *both* the spectral composition of the illuminant and the reflective properties of the surface illuminated. Somehow the eye seems to have an uncanny knack for extracting the latter information and disregarding the illuminant. In fact most people have the very definite perception that the color lies in the object (we speak of "colored objects") and not in the light from it. Speaking in a scientific sense we should rather say that the color is a sensation in our brain (it lies neither in the object nor its reflected light) that is stimulated by the light reflected from the object. Why then color constancy? Why do we receive the same sensation of red when the light reflected from the apple has a different spectral composition and is stimulating the red, green, and blue cones in different proportions in two of the above-mentioned situations? Of course color constancy is not absolute—if the illumination changes its spectral composition very drastically, almost all perceived object colors will change: The previously red apple will appear black under deep blue light. Also some surfaces have reflective properties such that they show perceptible color changes with relatively small changes in illumination; most people have had the experience of buying an article of clothing under artificial lighting only to have it change shade outdoors. Still the fact remains that most objects, under a wide range of illuminants, keep their perceived color.

Simultaneous contrast refers to the change in perceived color of some part of the field of view due to its **surround color.** The surround color always tends to induce its complementary color in the interior patch.[19, 20] This effect is most easily seen when the interior is a neutral gray, but even if it has its own color it becomes tinged with the surround complementary. This is almost the opposite problem from color constancy, in that the spectral composition of the light from the central patch has not changed from what it was without the surround, yet the perceived color has changed. One suggested explanation, in line with Young-Helmholtz theory, is that eye movements are bringing a complementary color afterimage of the surround onto the area in question, but this cannot explain the full effect in every case.[21]

9.4.4 Land's Experiments

In the 1950s a series of experiments in color vision were performed by the American scientist Edwin Land (see Chapter 5) which gave surprising results demonstrating both color constancy and simultaneous contrast. He made two simultaneous black-and-white transparencies of a scene through two different color filters, typically a green and a red.[22] An important factor in Land's experiments was that he used a natural scene with a variety of objects of different shapes and colors grouped together randomly. Almost all earlier investigations had used simple colored patches in groups of two or three, as in Fig. 9-14. The transparency through the red filter he

[19] *The Science of Color,* p. 155.

[20] Evans, *The Perception of Color,* p. 223.

[21] *The Science of Color,* p. 118.

[22] Edwin H. Land, "Experiments in Color Vision." *Scientific American,* May 1959.

called the **long** (for long wavelength) **record** and the one through the green the **short record.** Figure 9–15 shows an example of such a long and short record pair. Land projected the two transparencies together, while registering them on a screen, through their respective filters; that is the long record was projected through the red filter and the short record through the green filter. The full colors of the original scene were perceived with remarkable fidelity. The demonstration would not be remarkable if red, green, and blue records had been made and projected, as that is how Maxwell first demonstrated color photography almost a century before Land's work (see Chapter 11). But note that here the use of only two primaries has given nearly as good color reproduction, although the Young-Helmholtz theory would predict only greens, yellows, oranges, and reds. Yet an even bigger surprise awaits: if one of the filters is removed in the projection process, say the green, a wide range of colors is still reproduced, including pale greens! Remember that now only red and white light are being projected but the observer still sees many colors. Our three-pigment theory would predict just reds and pinks. Nor do the original record colors need to be as far apart as red and green—even two yellows of 579 nm and 599 nm have been used for the short and long records, giving a wide range of colors in the projected image. As the long and short wavelengths used for recording get closer together, the reproduced colors become more and more washed out. The experiment is showing color constancy in the extreme. Clearly some higher-order visual processes are taking place here.

Let us analyze Land's experiment more closely. Suppose there is a blue-green object in the natural scene somewhere. It will appear nearly black in the long record because the intervening red filter will block almost all wavelengths reflected from it. On the other hand it will appear quite light in the short record. When the long record is projected through a red filter, the image of our object will still be nearly black because the transparency transmits no light through that area. When the short record is projected with no filter, the image of our object will be filled with white light. Now because there is more red light in the areas around the image of our object, it should

Figure 9–15 Land's long and short records

appear blue-green by simultaneous contrast. However that cannot really be the whole explanation because if the short record is removed from the second projector so that it evenly covers the screen with white light, then all colors but the expected reds and pinks disappear.[23] This result is obtained even though the image of our object is still receiving about the same amounts of red and white light. The Land effects are therefore very difficult to understand in terms of the Young-Helmholtz theory, even if simultaneous contrast, which itself is not well explained within the theory, is invoked.

9.4.5 Retinex Theory

Eventually Land developed his own theory of color vision to explain these experiments. It deals with the processing of color information in the retina and visual cortex, and so has been named the **retinex theory.**[24, 25] The retinex theory does not exactly contradict the three-pigment theory, but it does probe much deeper into the brain's ability to produce colors in the final conscious picture. To simplify matters, the retinex theory first considers the ability of the human visual system to pick out differences in lightness of objects in black-and-white pictures regardless of the nature of the illumination, as well as the importance of edges or boundaries in that process. Figure 9–16 shows a pair of surfaces of different lightness which were unevenly illuminated from the left side so that the left surface received more light than the right. Everyone can immediately perceive that the left surface is darker than the right, even though the amount of light *reflected* from corresponding points on each is the same. How is the retinex able to pick out the object property (darker on the left, lighter on the right) from the illumination property (decreasing uniformly left to right)? Land's answer is that comparison across the boundary will accomplish the purpose. In demonstration, simply cover the boundary between the two surfaces in Fig. 9–16 with a pencil or other thin object: The perceived difference disappears and the two sides look identical, which they are as far as absolute amount of light in the retinal image is concerned. The retinex system automatically and subconsciously compares the amount of light from areas of each surface immediately adjacent to the boundary. Such a comparison eliminates any effect due to the uneven distribution of incident light. Furthermore, additional experiments have shown that the two surfaces do not have to be contiguous for such a process to work; the process can proceed from boundary to boundary across a number of intervening surfaces of varying lightness so that the viewer correctly perceives the relative reflective properties of two widely separated surfaces, all without conscious thought. Finally, to explain the facts of color vision, it is assumed that there are three such retinex systems, one for each of the primary colors, which operate in a similar manner to pick out the appropriate "light-

[23] Evans, *The Perception of Color,* p. 230.

[24] E. H. Land and J. J. McCann, *J. Opt. Soc. Am.* 61 (1971), 1.

[25] Edwin H. Land, "The Retinex Theory of Color Vision," *Scientific American,* December 1977.

Figure 9–16 Importance of edges in establishing lightness

ness'' of surfaces in red, green, and blue. Here we have a kind of blend of retinex theory with three-pigment theory.

This theory takes some giant strides toward explaining the facts of color constancy and Land's own color vision experiments. To return to our blue-green object in such an experiment, the theory predicts that the red-sensitive retinex system will assign very little red reflectivity to the object because it is darker that other areas in the red light from the long record; there is some red light in this area in the white light from the short record, but the relative intensities in the long record will control this red-sensitive response. The blue- and green-sensitive retinex systems will not respond to the red light from the long record but rather to the white light (which contains blue and green light) from the short record. In the short record the object is relatively light, so it is assigned a high reflectivity in blue and green. Thus the final conscious picture in the visual cortex has a blue-green object in it, just as it should have. If the short record were removed from the second projector, then there would be no lightness variations for the blue- and green-sensitive retinex systems to detect. In these circumstances only the *absolute* wavelengths can be detected: red in varying amounts mixed with white, giving rise to color sensations as predicted by the Young-Helmholtz theory.

Note the importance of a *natural image* in Land's experiments and theories. There must be a good deal of variation across the scene for the retinex to exclude the illumination effects and extract the reflectance information. We might say that in everyday color vision we are sensitive to variations in wavelength across the whole scene (which would be caused by the objects in the scene) rather than to absolute wavelength at any one point; Evans speaks of ''relative effective wavelength'' when describing the Land effects.[26] Such a viewpoint helps us to understand some other

[26] Evans, *The Perception of Color,* p. 231.

facts which are puzzling within the Young-Helmholtz theory. In the simple mixing of three spots of light of the primary colors, as in Fig. 9–14, many nonspectral colors cannot be produced: brown and metallic silver and gold, for example.[27] Yet when primary lights are mixed in a natural image on a color television screen, these colors, as well as seemingly all the colors we can perceive, are produced. Perhaps Land's greatest contribution was to emphasize the fact that the results of experiments with two or three spots of color will never be able to completely explain the rich variety of our everyday visual world.

QUESTIONS

1. What is a saccade?
2. Describe what happens when an image is stabilized on the retina.
3. How do we know a person can look at something and not see it?
4. Why does one line look longer in the Müller-Lyer illusion?
5. List five monocular depth cues.
6. List two methods of binocular depth perception.
7. Describe how 3-D movies worked.
8. Who invented the stereoscope?
9. What is meant by a stereo pair?
10. What is an anaglyph and how does it work?
11. Which mammals have color vision?
12. Who first proposed that there are three types of receptor in the eye, one for each color?
13. Who discovered color blindness?
14. What is the most common type of color blindness?
15. Which sex is more often color-blind?
16. What is a dichromat?
17. What causes protanopia?
18. Briefly describe the opponent color theory of color vision. Who originated it?
19. What color is the afterimage of a bright green light? Why?
20. What is meant by simultaneous contrast in color vision?
21. Briefly describe Land's color vision experiment.
22. In the retinex theory, how does the visual system determine relative reflectance values of objects within a natural image?
23. What colors cannot be produced in mixing three colored spots of light?

[27] Gregory, *Eye and Brain,* p. 125.

Part IV

Color

Chapter 10

Light and Color in Nature

We have already seen that the sensation of color is intimately connected with the wavelength of light (Chapters 2 and 9). In general, white light is a mixture of all the visible wavelengths in roughly equal amounts. Each individual wavelength of light gives rise to a sensation of one of the colors of the spectrum; that is, one of the colors of the rainbow or the continuum of light produced by passing sunlight through a prism. However, we also know that color vision is not as simple as a one-to-one correspondence between wavelength and color sensation; mixtures of wavelengths also give color sensations, both spectral and nonspectral. Also the state of adaptation of the eye and other colors in the field of view affect the color sensation at any one point. Nevertheless, for our purposes here we may at least associate color sensations with *variations* in the wavelengths of light across the field of view and inquire in this chapter about the physical causes of those variations.

10.1 NATURAL LIGHT

Our basic source of light in nature is the sun, which produces a continuous spectrum (all visible wavelengths) of light that has a peak in the middle. Figure 10–1 shows a graph of power from the sun versus wavelength, both above the atmosphere and at the earth's surface (sea level). You may note from the figure that the power at all wavelengths is less at the earth's surface because of atmospheric absorption. Also there are a number of sharp dips in the sea level curve, which are labeled with the chemical symbols for various compounds. These dips are due to absorption lines or

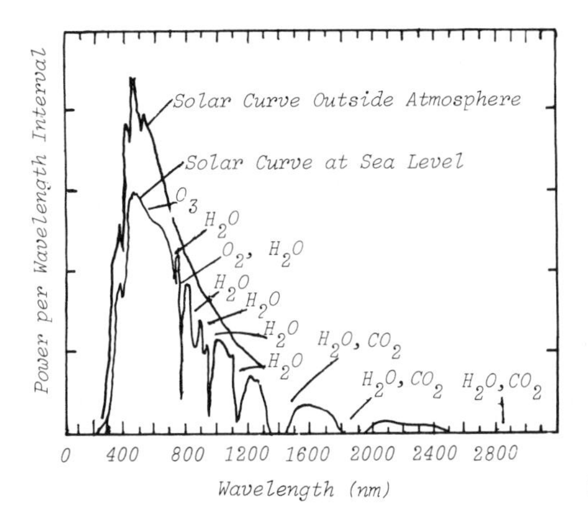

Figure 10–1 Solar spectrum

bands of various constituents of the atmosphere, and we shall consider them more closely a little later. For now let us just conclude from the sea-level curve that the sharp changes occur outside the visible region of the spectrum (400 to 700 nm), and that in the visible portion of sunlight all wavelengths are present in roughly equal amounts. In other words, as a first approximation we may think of sunlight as white, or uncolored. Actually, there is a sharper drop-off at the violet end of the spectrum than at the red end, which tends to make sunlight slightly yellowish, but in natural lighting objects are also illuminated by blue skylight so that the total tends more toward white.

If daylight illumination consists of all wavelengths in roughly equal amounts, where do the wavelength variations arise that lead to the tremendous array of colors that we see in nature? To answer that question we must look in some detail at the interaction of light with matter. We considered such interactions in Chapter 3 in discussing the Bohr model of the atom. There we had the very simple result that no interaction at all takes place unless the light is of precisely the frequency corresponding to the energy jump between electron levels (Chapter 3). But now we must look at matters a little more closely; nature is more subtle than that simple model.

10.2 FORCED OSCILLATORS

When an electromagnetic wave passes by an atom, it may be thought of as causing the outermost electrons of the atom to vibrate or oscillate slightly (in addition to their normal orbital motion), even though the frequency of the wave is not the correct one to cause a transition. Such a vibration or oscillation should not be surprising since an electromagnetic wave exerts an oscillating force on any charged particle that it passes. This viewpoint represents a very classical picture, and indeed some of the results we shall be interested in were derived before there was a quantum theory. It is just this very small oscillation, less than 10^{-17} m in amplitude,[1] that gives rise to many beautiful lighting and color effects in nature.

There exists a completely worked out classical theory for the motion of an oscillator subjected to a periodic oscillating force. This theory may be applied directly to our model of an electron bound in an atom (the oscillator) acted upon by an electromagnetic wave (the periodic force). In fact we shall even enlarge our viewpoint to consider the vibrations of a whole atom bound in a molecule and acted upon by an electromagnetic wave. Any oscillator is most simply thought of as a mass on a spring; for our electron in an atom, the mass is the electronic mass and the spring represents the force that holds it in the atom, while for the atom in a molecule, the mass is the atomic mass and the spring represents the force that holds it in the molecule. A mass on a spring (and any oscillator it represents) always has certain **natural frequencies of oscillation;** these are the frequencies of oscillation that the mass has if it is disturbed slightly and then allowed to move on its own. The lowest and most basic such natural frequency, which we shall call f_0, depends on the mass and the strength of the spring. Generally, the smaller the oscillating mass and the stronger the spring, the higher is f_0. Thus, if a mass on a spring is displaced slightly from its rest position and released, it will vibrate with a certain natural frequency. If the same experiment is repeated with a larger mass on the same spring, the mass will vibrate with a lower natural frequency. Or if the original spring is replaced with a stronger one, the original mass will vibrate with a higher natural frequency.

We may make a somewhat awkward but useful combination of classical and quantum theories by identifying the natural frequencies of our oscillators with the frequencies required for quantum jumps, as given by Eq. (3) in Chapter 3. Even the atom bound in a molecule has certain energy states that it can be in and shows quantum jumps between those states. So we can define a set of natural frequencies using Eq. (3) in that case also.

Now when a periodic force is applied to a mass on a spring, the theory tells us that the mass will vibrate at the same frequency as the applied force. But it would prefer to oscillate at its own natural frequency, so the amplitude of vibration depends on how close the applied frequency f is to that natural frequency f_0. Figure 10–2 shows this preference as a plot of the amplitude of vibration versus f/f_0. Several comments should be made about Fig. 10–2 before proceeding to apply it to our problem. First, the value of the quantity f/f_0 on the horizontal axis tells us how close the applied frequency is to the natural frequency of the oscillator; $f/f_0 = 1$ when the two are equal, which condition is often called **resonance.** Sometimes the natural frequencies are also called **resonant frequencies.** Second, you should be aware that the theory which gives the curve of Fig. 10–2 also assumes that there is some type of frictional force (often called **damping**) acting on the oscillator and converting some of the energy into other forms such as heat. This assumption is always a good one in any real situation (even one involving atoms). We may and do assume that the damping force is small, but some such force is needed to keep from getting infinite results at

[1] Victor F. Weisskopf, "How Light Interacts with Matter," *Scientific American,* September 1968.

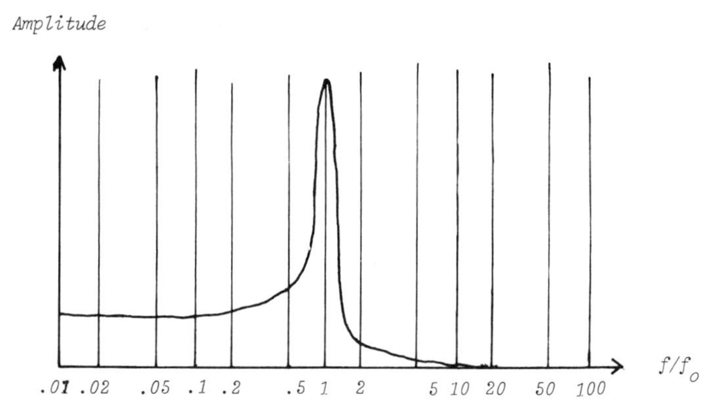

Figure 10–2 Amplitude of forced oscillator

resonance. The details of the curve shape, such as the height of the resonance peak compared with the level portion on the left, depend on how large a damping force is assumed.

The most obvious piece of information to be learned from the graph is that the amplitude of vibration of the oscillator is greatest at resonance, when $f = f_0$. Even without any elaborate theory we could have guessed that any oscillator would respond the most when a force pushes it at just the frequency at which it would like to vibrate on its own. At resonance the oscillator picks up energy very readily from the applied force. One example of resonance with which you may already be familiar is the case of pushing someone on a swing. In this case you push in the same direction each time it comes back to the same position in its oscillation, matching the frequency of your applied force to the natural frequency of the swing; a much more effective procedure than trying to push forward each time when the swing is in various different parts of its cycle. The graph also shows that when f is much less than f_0 ($f/f_0 << 1$), then the amplitude of vibration is small compared with resonance but rather independent of f; that is, the amplitude changes almost not at all as f/f_0 goes down from 0.1 to 0.01 and smaller. Finally, we note that for f much greater than f_0 ($f/f_0 >> 1$) the amplitude goes rapidly to zero; an oscillator hardly responds at all to applied forces with frequencies more than 10 times the resonant frequency.

The theory actually tells us more than is shown in the graph. It tells us the phase of the oscillator's motion relative to that of the applied force. Thus, for $f/f_0 << 1$ the oscillator and applied force are in phase, which means that when the force on the oscillator reaches a maximum in one direction, the oscillator has a maximum displacement from its rest position in that same direction. At resonance, $f/f_0 = 1$, they are said to be 90° out of phase, which means that when the force reaches a maximum the oscillator is passing through its rest position (zero displacement). Finally for $f/f_0 >> 1$ they are said to be 180° out of phase, or opposite in phase, which means that when the force reaches a maximum in one direction the oscillator has a maximum displacement in the opposite direction. This phase information will also prove useful.

10.3 RAYLEIGH SCATTERING

What does all this mean in terms of light and matter? Suppose the oscillator to be an electron bound in an atom or an atom bound in a molecule and initially in its lowest possible energy state (ground state). For most ordinary substances the resonant frequency (or frequency corresponding to the first quantum jump) for an electron in an atom is in the ultraviolet, well above visible frequencies. For an atom in a molecule, the resonant frequency is quite likely to be in the infrared, well below visible frequencies. This result is reasonable even from a classical point of view in which the particles are viewed as masses on springs. The electron is very light (small mass) and bound in the atom rather strongly (strong spring), which leads to a high natural frequency of oscillation. An atom is much heavier (large mass) and bound by weaker forces in a molecule (weak spring), which leads to a comparatively low resonant frequency. Ultraviolet radiation has higher frequencies than light while infrared has lower frequencies. Let the atoms and molecules be those of our atmosphere: mostly nitrogen, N_2, and oxygen, O_2 (O_3 when in the form of ozone), with smaller amounts of carbon dioxide, CO_2, and water, H_2O. The sharp dips in the curve of Fig. 10–1 show some of those resonant frequencies (or corresponding wavelengths) outside the visible spectrum. When sunlight strikes these molecules of gas, their response to the visible part of the spectrum is small but not wholly negligible. For the infrared oscillators (the atoms in the molecules) we have $f/f_0 >> 1$, and as we noted earlier the amplitude of vibration is essentially zero; that is, the infrared oscillators do not absorb any energy from the visible wavelengths to speak of. But for the ultraviolet oscillators (the electrons in the atoms) we have $f/f_0 << 1$, and there is a small amplitude of vibration that is about the same for all visible frequencies. That means that the electrons in the atoms absorb energy from all the visible wavelengths and oscillate with about the same amplitude no matter which frequency of vibration in the visible they pick up. The energy per second, or power, which they absorb is for the most part reradiated in all directions, since an oscillating electric charge radiates electromagnetic waves at its frequency of oscillation. Now the power absorbed and emitted by such an oscillating charge depends on both its amplitude and its frequency, increasing as either one increases. Again it should be reasonable to expect that more power is required to force an oscillator to vibrate through a larger amplitude, and also that more vibrations per second (a higher frequency) imply more energy per second. All the ultraviolet oscillators of the atmosphere vibrating at visible frequencies have the same amplitude, but those with higher frequencies in the visible (blue) are absorbing and reradiating more power. And that is why the sky appears blue! Sunlight comes in from one direction and blue light is scattered out in all directions by the electrons in

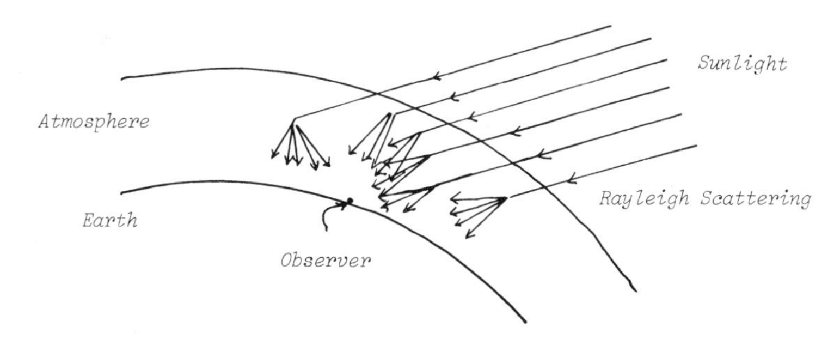

Figure 10–3 Atmospheric scattering

the atoms of the atmosphere, so that an observer looking toward the sky in any direction sees this blue light. Figure 10–3 illustrates this effect.

This explanation for the blue color of the sky was first worked out classically over a century ago by the great English physicist John William Strutt,[2] better known as Lord Rayleigh (1842–1919). For this reason the effect is still known as **Rayleigh scattering.** In general, the term *scattering* is used whenever light is absorbed and very quickly reemitted in all directions. Rayleigh scattering accounts for many colors in nature beside the blue sky. Most notably, the red of sunset or sunrise is simply the color left in direct sunlight after a long trip through the atmosphere has scattered out the higher frequencies (shorter wavelengths). Figure 10–4 illustrates the longer atmospheric path for sunlight reaching an observer at sunrise or sunset.

Sunrises and sunsets can be particularly spectacular when there are many small particles suspended in the air (dust or even human-caused pollutant particles) because each one acts as an additional scatterer. However, the particles must be smaller than one wavelength to contribute to the color; we shall return to this point shortly. Thus, unusually vivid sunrises and sunsets are common after large volcanic eruptions have spread fine ash into the atmosphere. Bluish hazes seen in the daytime are attributable to Rayleigh scattering from small suspended particles. When looking at distant mountains you can often see a bluish haze against the darker part of the mountains, whereas the snow-covered portions may look yellowish. This effect, called **aerial**

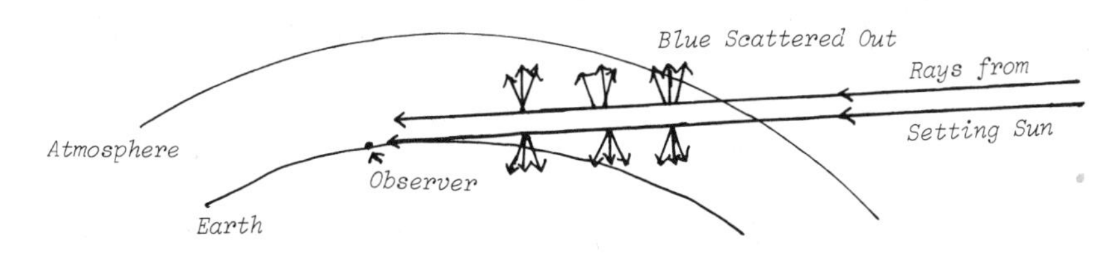

Figure 10–4 Sunrise or sunset

[2]Earl J. McCartney, *Optics of the Atmosphere: Scattering by Molecules and Particles* (New York: John Wiley & Sons, Inc., 1976), p. 177.

perspective, has often been used by artists to achieve depth. Where there is little background light (the dark portions of the mountain) you can see the Rayleigh-scattered light (blue haze), but where there is a good deal of background light, you see that with some of the blue scattered out (the yellowish snow). Since only a little blue light is scattered from moderate path lengths, you usually need a dark background to see the scattered light. Thus, blue eyes of humans are blue not because the irises have a blue pigment, but rather because they have a number of small scattering particles in front, giving Rayleigh scattering seen against the dark of the rear layers of the iris. Even the blue sky is seen against the black of outer space.

10.4 WHITE LIGHT SCATTERING

In order to have Rayleigh scattering, with its preference for blue, the scattering particles must be smaller than a wavelength of light and independent in their action. Such conditions are easily satisified by the molecules of a gas, whose dimensions are only about one-thousandth of a wavelength and which interact only when they collide. Small particles suspended in some medium also qualify. But in a solid particle or liquid droplet that approaches the size of a wavelength of light, we must take into account additional effects due to the influence of the separate atoms upon each other. The forces between atoms in a solid or a liquid tend to make them act in unison so that the light they absorb and reradiate shows interference effects. In fact, the effects tend toward constructive interference in the forward direction and cancellation everywhere else in the bulk of the material, which just leads to the refracted wave with which we are already familiar (see Chapter 4). Near the surface of the material in a layer about one-half wavelength deep, the atomic oscillators contribute to a backward wave also, the reflected light. For larger wavelengths (smaller frequencies) more atoms contribute to the constructive interference, and it turns out that this effect just cancels out the individual oscillators' preferential absorption and reradiation of shorter wavelengths (high frequencies). So, if the incoming light is white, the reflected and refracted light is white (neglecting dispersion, an effect to be considered later). If many particles which are small but still larger than a wavelength are reflecting, we still often speak of the light as *scattered,* since it is reflected in all directions by the collection of particles.

We can now understand in a rough manner what happens to the optical properties of a small particle of a simple substance, such as a water droplet, as its size changes. If it starts out smaller than a wavelength of light, it exhibits Rayleigh scattering, preferentially scattering the blue. As it grows to larger than a wavelength of light, the blue preference decreases more and more until the particle is scattering all wavelengths equally; that is, it is scattering white light. Now it may be a water droplet in a cloud or a fog which appears white. Another simple and striking example of this scattering dependence on particle size is given by a lighted cigarette. The smoke from the burning tip has a definite bluish cast (most easily seen when viewed against a dark background), a sure indication that the smoke particles are mostly smaller than a wavelength. However, the smoke from the user's end is much whiter, indicating that the particles are larger than a wavelength—they have grown by picking up tars and water as they pass through the length of the cigarette.

Of course a large enough chunk of water looks clear rather than opaque white like a cloud. Even each small droplet in a cloud would look clear when seen by itself because most of the incident light is transmitted through in the refracted wave and only a small fraction is reflected. However, when many small droplets are seen in aggregate, the small fractions reflected at each surface of each particle add up to a significant total. The same thing is true for any clear substance: when divided into a mass of small particles, it will appear opaque white. Thus, each crystal of sugar or salt is relatively transparent, but a pile of either is white. Snow crystals are clear (like ice cubes) but everyone knows snow on the ground is white (as is finely crushed ice). Even this white paper is really composed of many small, relatively clear cellulose fibers stacked on top of one another.

10.5 DISPERSION

Since we have been discussing the sky and water droplets, this would seem to be a good place to consider one of the most spectacular color displays in nature, the **rainbow.** The display of colors in the rainbow is due to the same process that breaks up white light into colors in a prism. This phenomenon is called **dispersion** and simply means that not all frequencies (or wavelengths) travel with equal speed in the medium. Referring to our earlier discussion of refraction (Chapter 4), we could also characterize dispersion by saying that the index of refraction of the medium is slightly different for different frequencies. All transparent substances show some dispersion in the visible and, if uncolored, have a higher index of refraction for the violet than for the red.[3] This difference implies that higher visible frequencies are slowed down more by the medium than are lower frequencies. For example in silicate flint glass, which has large dispersion, the refractive index n is 1.66 at 400 nm and 1.61 at 700 nm.[4] For water, whose dispersion is responsible for the rainbow, the effect is smaller: n is 1.339 at 400 nm and 1.331 at 700 nm.[5] It is remarkable that such a small change (less than 1 percent) in one property of water across the visible spectrum can lead to such a vivid display of color in nature.

Dispersion can also be explained by considering the response of atomic oscillators to the electromagnetic waves. We said before that the tendency of more

[3] Frances A. Jenkins and Harvey E. White, *Fundamentals of Optics,* 4th ed. (New York: McGraw-Hill Book Company, 1976), pp. 476–478.

[4] Francis Weston Sears, *Optics,* 3rd ed. (Cambridge, Mass.: Addison-Wesley Publishing Co., Inc., 1956), p. 49.

[5] Leo Levi, *Applied Optics: A Guide to Optical System Design,* vol. 2 (New York: John Wiley & Sons, Inc., 1980), p. 980.

oscillators to act in unison for lower frequencies just cancels the individual oscillators' preference for absorbing and reradiating power at higher frequencies in solid or liquid particles that are themselves larger than a wavelength. This cancellation of effects, however, is only exact if the amplitude of vibration is exactly the same for all the visible frequencies. Since we are only concerned here with the oscillation of electrons in atoms (resonances in the ultraviolet), we are looking at the left side of the curve in Fig. 10–2. When we look at the fine details, we can see that the amplitude of vibration at violet frequencies should be a tiny bit larger than those at red frequencies because the former are closer to the resonant ultraviolet frequency. Thus, in the total effect, the whole refracted wave, there should still be some slightly increased tendency for the material to act on the higher visible frequencies. This action shows up as a higher index of refraction, or slower speed, for violet. The slowing is due to the absorption and reemission by the atoms within the material: the more such absorption takes place, the slower the travel of light through the medium because of the slight time delays before reradiation. In a way, one could think of violet light as a kind of "local bus," which makes many stops compared with red light, which is more of an "express" and accordingly makes the trip faster.

Since each different wavelength of light has a different refractive index in some transparent medium, each will bend at some slightly different angle upon entering the medium; violet will be bent the most and red the least, with the other spectral colors in between. If the boundary of the medium at which the light exits is not parallel to that at which it entered, then the colors will not be brought back together upon leaving the medium and a separation of the colors by angle will result. Figure 10–5 shows the effect for a simple glass prism.

10.5.1 Rainbows

To form a rainbow we need droplets of water (presumed spherical) with direct sunlight on them. The primary rainbow is formed by light that enters on one side, reflects off the back of the drop, and then exits nearer to where it entered. Figure 10–6 illustrates with a single drop and eight numbered rays of light of the *same* wavelength, say green. One can see from the figure that the rays tend to pile up at the one angle of 41° with respect to their original direction. This result follows from a detailed analysis of the reflection and refraction of light in a spherical volume of water. But now we remember that each different wavelength of light has a different index of refraction in water: violet bends more and red less at each of the two refractions. So violet light tends to pile up, or be most intense, at an angle of 40°, while red does the same at 42°. An observer on the ground sees blue light coming from every

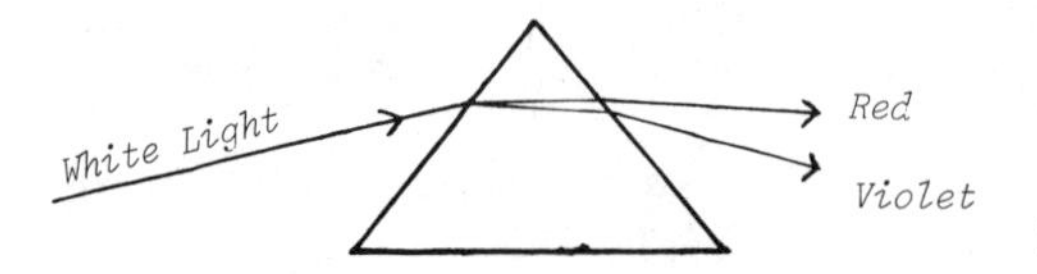

Figure 10–5 Dispersion in a prism

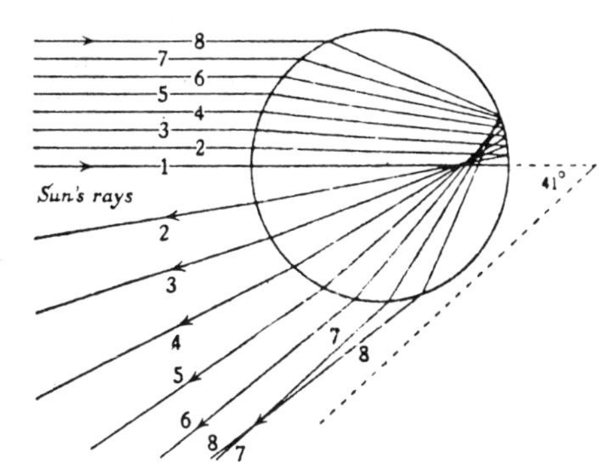

Figure 10–6 Green light in raindrop (From Sir William Bragg, *The Universe of Light,* New York: Dover Publications, Inc., 1959).

raindrop along a line making an angle of 40° with respect to the sun direction, red light from every drop along a line at a 42° angle, and the other spectral colors along lines at intermediate angles, as illustrated in Fig. 10–7.

To understand fully the geometry of the rainbow, one should realize that Fig. 10–7 can be rotated around the line *OS* to give a three-dimensional view. Then the lines *OR* and *OB* sweep out cones with an apex at *O,* and the colors seem to be coming from circular arcs in the sky rather than just one direction. Furthermore, although it is convenient for the figure, the line *OS* between the sun and the observer does not have to be horizontal. The point is that the rainbow is seen to be 40° to 42° off this line and centered on it no matter what direction the line takes.

From Fig. 10–7 several features of the rainbow can be explained. First we can see that the order of colors progresses from red on the outside to blue on the inside of the bow. Note that the rainbow is always seen on the opposite side of the observer from the sun, but it is not localized in space. All the light of a single color is coming from one *angle* rather than one *position* in space. This nonlocalized property is why you can never reach the mythical pot of gold at the end of the rainbow: if you move in that direction, new raindrops farther away than the original ones will be sending you the colored light.

The geometry of Fig. 10–7 places some limitations on the viewing of a rainbow. Obviously, direct sunlight must be striking some rain beyond the observer. Since the

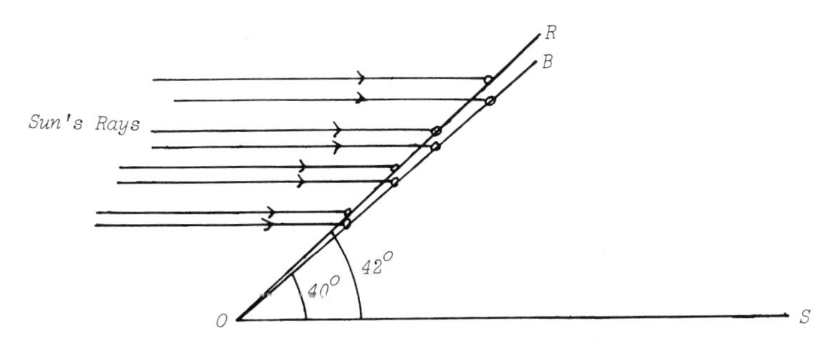

Figure 10–7 Formation of the rainbow

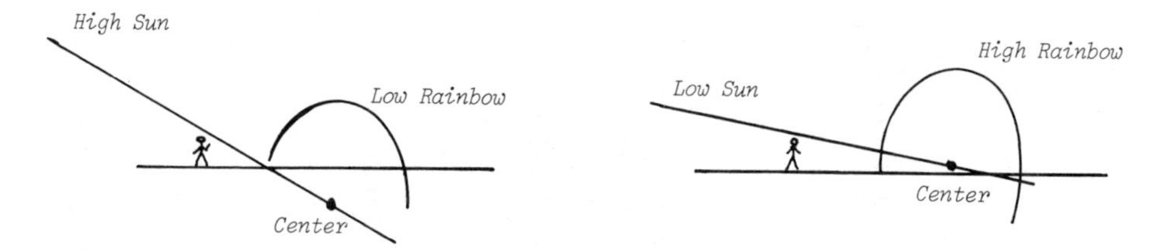

Figure 10–8 Sun-rainbow relation

rainbow is centered on the line from the sun to the observer, a high sun means a low rainbow and vice versa. Figure 10–8 illustrates. If the sun is more than 41° above the horizon, then the rainbow cannot be seen by a ground observer because it is all below the horizon. On the other hand, an observer on a mountaintop or in an airplane may see the entire circular arc of colors, as shown in Fig. 10–9.

These relations show why the most vivid rainbows are often seen after a late afternoon thunderstorm when the sun sinks quite low in the sky. Finally, a careful study of Figs. 10–6 and 10–7 reveals that the sky inside the rainbow should be brighter than the sky just outside. This result is due to the fact that raindrops below the line *OB* of Fig. 10–7 will be sending rays 2, 3, 4, or 5 toward the observer at *O* (think of placing the raindrop of Fig. 10–6 somewhere below *OB*). Since all wavelengths will be mixed in this light, it should just appear white. However, because no light comes out of the raindrops after one reflection at an angle greater than 42° from the original direction, drops above the line *OR* of Fig. 10–7 send no light to the observer at *O* (think of placing the raindrop of Fig. 10–6 somewhere above *OR*). Thus no light at all comes from the sky just outside the rainbow.

Actually, if you look far enough outside the primary rainbow you may see light that has been reflected twice before leaving the raindrops, forming what is known as a **secondary rainbow.** Figure 10–10 shows an example of a ray path for the secondary rainbow. The secondary rainbow is always less intense than the primary bow because less light will make the two reflections necessary. The extra reflection is also responsible for the reversal of color order in the secondary bow (red on the inside to blue on the outside) and for the reversal of light and dark areas of the sky (dark inside the bow and light outside). Thus when both the primary and secondary rainbows are visible,

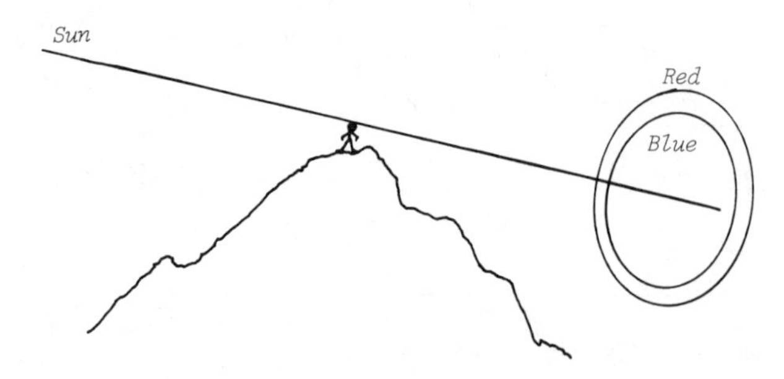

Figure 10–9 Rainbow for an observer above horizon

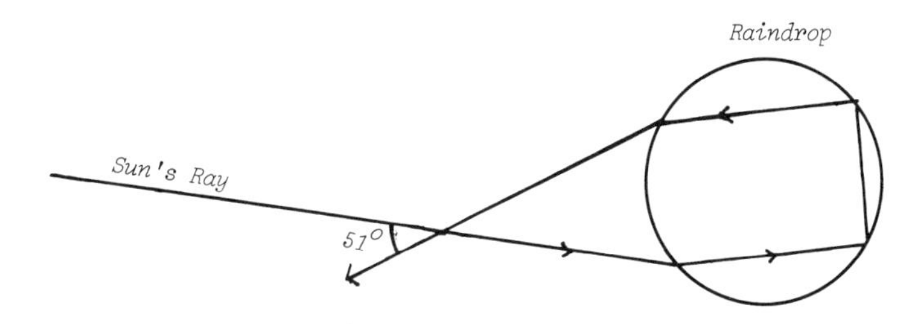

Figure 10–10 Ray path for secondary rainbow

the sky is notably darker *between* them than either inside the primary or outside the secondary. Figure 10–11 shows the geometry involved in formation of both bows, with the secondary always at an angle of 51° with respect to the sun-observer line.

When a rainbow forms with a very bright sun, you may sometimes see a number of still dimmer arcs of washed out violet-pink and green inside the primary rainbow. These arcs are called the **supernumerary bows,** and they cannot be explained by simple geometrical optics; they are due to interference effects. The details of their appearance can actually give information about the size of the raindrops responsible. For our purposes, we only note that the supernumerary bows are most vivid and numerous when the raindrops are larger, about 1 to 2 mm in diameter.[6]

10.5.2 Halos

Various other kinds of **halos** seen in the sky, not as frequent but often quite striking, have explanations similar to the rainbow. One of the commonest and easiest to explain is the 22° halo around the sun which usually appears when high, thin, cirrostratus clouds turn the sky milky white. This appears as a ring of light around the sun, 22° away from it in the sky, with a sharply defined inner edge of red, followed by yellow, green, and white tending to blue. The colors are quite unsaturated compared

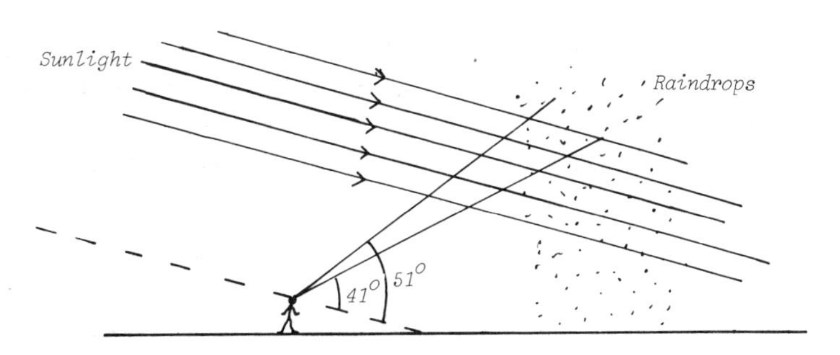

Figure 10–11 Formation of primary and secondary rainbows

[6] M. Minnaert, *The Nature of Light & Colour in the Open Air* (New York: Dover Publications, Inc., 1954), p. 178.

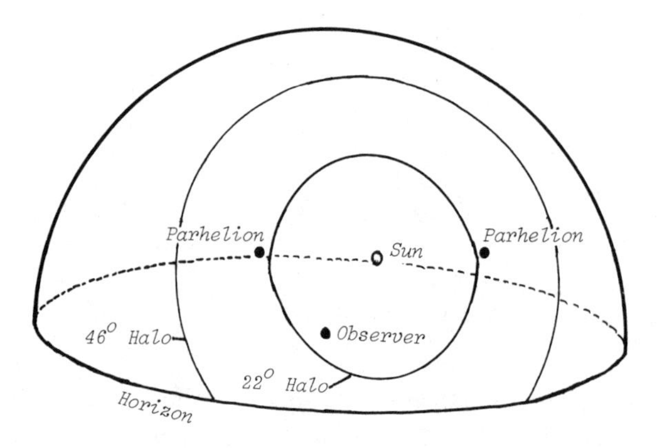

Figure 10–12 Lights in the sky

with those of the rainbow. Although 22° may not sound very large, the halo may seem quite enormous when actually seen in the sky; if it is really the 22° halo, your spread hand, thumb to little finger, at the end of your outstretched arm should just about reach from sun to halo.[7] Figure 10–12 shows the geometry involved.

The high clouds that are necessary for the 22° halo are composed of ice crystals rather than water droplets, and it is refraction and dispersion in those ice crystals that lead to the halo. The hexagonal symmetry of the ice crystals is responsible for the 22° angle because, with n being 1.31 for ice, the rays which strike the side of the crystal at different places tend to "pile up" (as in the water drop for the rainbow) at an angle of 22° with respect to their original direction. Figure 10–13 shows a ray which is just entering at the right position to give the 22° angle, which is sometimes called the **angle of minimum deviation.**

Now to get the complete circular halo, there must be randomly oriented ice crystals in the cloud so that those that are at the correct orientation around a circle in a plane perpendicular to the line of sight can all send light to the observer. Figure 10–14 illustrates.

The colors arise because the angle of minimum deviation is slightly different for the different wavelengths. In other words, ice also shows dispersion which is actually greater than that of liquid water: for ice $n = 1.317$ in the violet and $n = 1.306$ in the red.[8] Although there are obvious similarities between the rainbow and the 22° halo, it is instructive to note the differences. Because there is no reflection involved in the formation of the halo but there is in that of the primary rainbow, the order of the colors

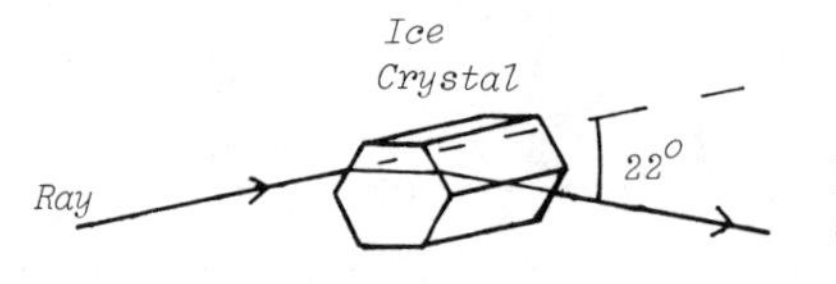

Figure 10–13 Ray through side of ice crystal

[7] Ibid., p. 190.

[8] Jenkins and White, *Fundamentals of Optics,* p. 476.

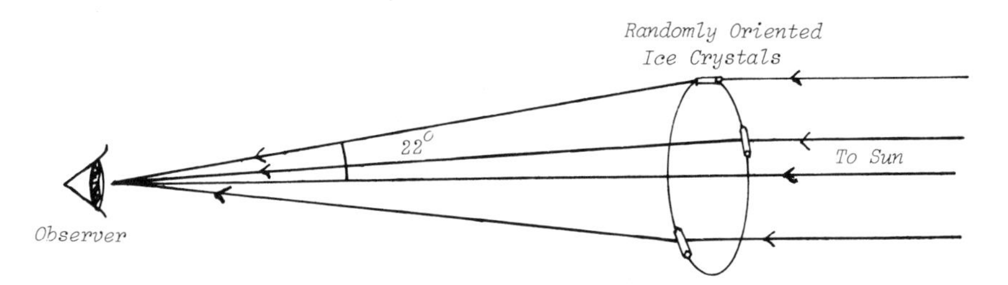

Figure 10–14 Formation of 22° halo

is opposite in the two cases: red is inside and blue outside in the halo. The same reflection also accounts for the fact that the rainbow is seen generally opposite the sun while the halo is seen in the direction of the sun.

We should also note that the 22° halo is sometimes visible around the moon, but much smaller colored rings of light known as coronas are also often seen around the moon. These coronas are different in appearance and origin, and we shall return to them later. There even exists at least one report of simultaneous 22° halos seen around the setting sun and rising full moon.

One recent work[9] disputes the classical assumption that halos are formed by *randomly* oriented ice crystals, but finds their origin rather in partially aligned ice crystals, 12 to 40 μm in diameter. Regardless of that disagreement, scientists do agree that many other optical sky phenomena are caused by such aligned ice crystals. One example, commonly seen with the complete or partial 22° halo, is the **mock sun** (also called **parhelion** and **sun dog**).[10] Often two such mock suns are formed, one on each side of the real sun at the same altitude, appearing as bright patches of light on or near the 22° halo (nearer when the sun is lower). These are the points marked parhelia in Fig. 10–12. One or both mock suns may be visible even when the 22° halo is not. It seems that some ice crystals in clouds form as hexagonal plates on top of hexagonal rods, much like tiny umbrellas, and so tend to align themselves with their long dimension vertical. Figure 10–15 illustrates. Referring to Fig. 10–14, one can see that if there are more of these vertical crystals that other orientations, the sides of the 22° halo will show up more brightly. If only the vertically aligned crystals are present, then only the sides of the halo will show up, so that only two mock suns will be visi-

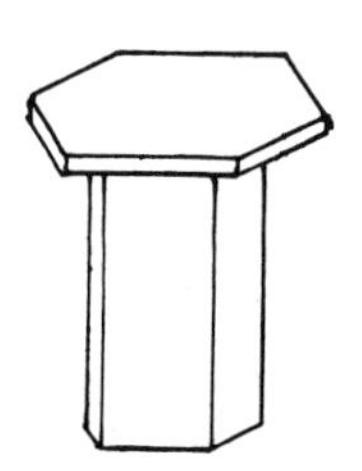

Figure 10–15 Ice crystals responsible for parhelia

[9] Alistair B. Fraser, *J. Opt. Soc. Am.*, 69, (1979) 1112.

[10] David K. Lynch, "Atmospheric Haloes," *Scientific American*, April 1978.

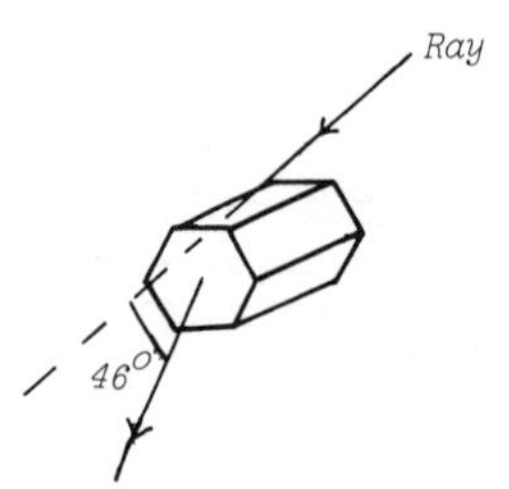

Figure 10–16 Ray through end face

ble. Of course, when the appropriate crystals are only on one side of the sun only one parhelion is seen.

Also shown in Fig. 10–12 is a larger circle of light known as the **46° halo.** As you might have guessed, this halo is always seen 46° away from the sun. The colors are in the same order as the 22° halo and the same sort of ice crystals give rise to both halos. However, the 46° angle represents the angle of minimum deviation for rays that enter one side of the crystal and exit through the end face at 90° to the side, as shown in Fig. 10–16. Thus the formation of the 46° halo depends on the presence of ice crystals of the correct orientation so that these rays reach the observer. With enough crystals randomly oriented, the ones required should be present and the 46° halo should appear. Many other optical effects in the sky can be caused by the ice crystals in clouds, and the interested reader is referred to Lynch and Minnaert.

10.6 CORONAS AND THE GLORY

We briefly mentioned earlier the smaller rings of colored light, called **coronas,** which are often seen around the moon through thin clouds. These coronas can be easily distinguished from the halos we have discussed by two features: they are much smaller, being only a few degrees in radius, and the order of colors is different, with blue on the inside of the coronas. The sun can also have coronas around it, but they are more difficult to see because one has to look too directly at the dazzling sun itself. Coronas are thought to be caused by interference of light waves that strike a small water droplet or ice crystal and pass around the sides, spreading out behind it in the process called diffraction (see Chapter 2). A very vivid corona indicates a uniform diffracting particle size of about 0.015 mm.[11]

Before leaving striking optical effects in the sky, there is one more ring of light worth discussing because of the influence it may have had on religious art. This much rarer colored ring of light is called the **glory.** Originally it was seen by observers at the top of a hill or mountain when the sun was low and the observer's shadow was cast

[11] Minnaert, *Light & Colour in the Open Air,* p. 218.

upon a layer of mist. Now it more often occurs for observers in airplanes. It appears as a small ring of colored light (red outside to bluish inside)[12] surrounding the shadows of the observer's head and only a few degrees in radius.[13] The cause of the glory is still the subject of scientific investigation, but it certainly is in the realm of wave optics since geometrical optics gives no explanation. Recent work suggests that it stems from backscattering from mist droplets with diameters several hundred to several thousand times the wavelength of light.[14] We know that in the past mountaintops were places of retreat and contemplation for religious mystics. It is altogether possible that the observation by one or more such mystics of their own looming shadow, accompanied by a vivid glory, on a fog bank in early morning or evening, was interpreted as a divine apparition and subsequently led to the artistic tradition of halos around heads of holy persons. In this respect it is interesting to note that if several people are present, each sees all the shadows, but sees the glory around his or her own head shadow only.[15] Lest this one rather facile historical theory be too easily accepted, it is necessary to warn that there are other natural lighting effects which could have led to the same result, and we shall discuss one of them later.

10.7 THIN FILM COLORS

We have already moved into a realm where many color effects have required wave interference for even a basic explanation: supernumerary rainbows, coronas, and the glory. A number of other beautiful colors in nature that we do not have to look skyward to see are also explained by interference. In particular, thin layers of transparent or translucent materials can give rise to these interference colors. One example that you have probably seen is a thin film of oil on top of water. When a light rain has fallen on a blacktop parking lot, you can often see irregular circular patches of color glistening on the wet pavement. These irridescent patches take the form of colored rings and change in appearance with changing viewer angle. The color comes from interference between waves that have been reflected at the top and bottom of a very thin film of oil dripped from automobiles and now floating spread out on a layer of water. Figure 10–17 shows a highly magnified view of such an oil film.

Light waves reflected from the bottom surface can interfere destructively with waves reflected from the top surface if the path difference is an odd number of half wavelengths and constructively if it is a whole number of wavelengths. The path dif-

[12] L. Larmore and F. F. Hall, "Optics for the Airborne Observer," *SPIE Journal,* March 1971, p. 91.

[13] Ibid., p. 89.

[14] H. M. Nussenzveig, *J. Opt. Soc. Am.,* 69 (1979) 1068.

[15] Minnaert, *Light & Colour in the Open Air,* p. 224.

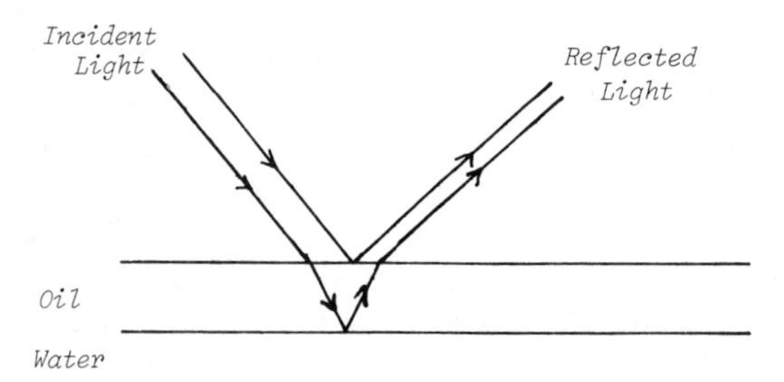

Figure 10–17 Interference colors in oil film

ference (the extra distance traveled by the wave off the bottom) is just twice the film thickness for light entering and leaving nearly perpendicularly. Generally, the oil spreads into a circular shape, thicker in the middle than at the edges. This shape leads to light and dark rings (areas of constructive and destructive interference), much like a Gabor zone plate (see Chapter 7) for any one wavelength. With white light entering the film we see the superposition of light and dark rings, having different radii for different wavelengths, for all the colors together, leading to a complicated series of colors as the final result. If we look at the patch from an oblique angle, we see light that entered at an oblique angle and there will be a greater path difference, leading to a rearrangement of color. Other substances, such as turpentine on water, will produce the same effect, but apparently only *used* motor oil is effective.[16] Since only a small fraction of the light is reflected at either surface, the colors are rather dim and require a dark background for good visibility.

The iridescent colors in a soap bubble represent another good example of interference colors. Here the interfering waves are those reflected from the inner and outer air-water interfaces. The opalescent colors of pearls and some seashells (such as abalone shells) are also due to interference. In these cases multiple translucent layers, deposited in a natural growth process, give the multiple reflections for interference. For clear multiple layers the interference colors can become quite vivid. In bird feathers, butterfly wings, and beetle shells, interference colors can be combined with pigments to produce spectacular results.[17]

10.8 PIGMENTS IN NATURE

The mention of pigments brings us to a fourth process for producing color from white light—**absorption.** By considering absorption last we do not mean to imply that

[16] Ibid., p. 210.

[17] W. D. Wright, *The Rays are not Coloured: Essays on the Science of Vision and Colour* (New York: American Elsevier Publishing Company, Inc., 1967), pp. 8–16.

it is less important than the other three, namely scattering, dispersion, and interference. In fact, division into four processes is essentially a human artifact for convenience in explanation and calculation. Nature is a unity, and all color arises basically from the interaction of electromagnetic radiation with matter. Far from being the least important, absorption accounts for the colors of pigments, most of the colors we see around us. By absorption we simply mean that the substance strongly absorbs some visible wavelengths and converts the electromagnetic wave energy into other forms. Referring back to our earlier discussion of forced oscillators, one may realize that we are now considering substances with resonant frequencies in the visible (the only way visible frequences can be strongly absorbed). But we said then that most simple atoms and molecules have resonances outside the visible, in the ultraviolet or infrared. However the organic molecules found in nature are not usually simple. Often they are made up of dozens or even hundreds of atoms, and sometimes electrons are free to move throughout the molecule. Then the electronic resonant frequency can fall between the infrared and ultraviolet, that is, in the visible. We have a kind of intermediate situation with molecular forces (weak spring) and electron oscillator (small mass), leading to an intermediate resonant frequency. This is the case with some of our most common natural pigments.

When you view a landscape in sunlight in spring or summer, you see two colors that predominate, the blue of the sky, already discussed, and the green of living vegetation. It is an interesting fact that you can heighten the color sensations by turning your back to the scene and then bending over and viewing the landscape upside down between your legs. Some have suggested that the effect is due to blood rushing to the head.[18] In any case, the green that seems to be everywhere on earth is due to absorption by molecules of **chlorophyll** in living plants. That chlorophyll is not a simple molecule can be seen from the fact that it contains 55 atoms of carbon, 72 of hydrogen, 5 of oxygen, 4 of nitrogen, and 1 of magnesium, 137 atoms in all. This molecule has a broad absorption band in the orange-red part of the visible spectrum and a smaller absorption band in the violet. Note that the resonant frequency is *not* at the color that the substance appears. The color we see reflected from pigments is what is left in white light after the resonant frequencies have been absorbed. After a band of red-orange and a few other frequencies have been absorbed by chlorophyll, the remainder of the sunlight reflected to an observer looks green.

Another common organic molecule that often gives rise to color in nature is **carotene,** composed of 40 carbon and 56 hydrogen atoms. There are different forms of the carotene molecule, corresponding to different three-dimensional arrangements of atoms, so we often speak of *carotenes* in the plural. The name is due to the fact that it was first extracted from carrots, which derive their color from it. Obviously carotenes must absorb primarily in the violet to green in order to give the reflected colors yellow to red. Carotenes are also present in animal substances, and in

[18] Minnaert, *Light & Colour in the Open Air,* p. 113.

fact vitamin A is produced from carotenes in the human liver. Thus we find that the yellows of egg yolks and of butterfat in milk are due to the presence of carotenes. They are also responsible, along with a similar molecule, for many of the yellows to reds of flowers and autumn leaves. In the latter case, the yellow and orange pigments were present even in the spring and summer, but they were masked by the effect of chlorophyll. In the fall the chlorophyll disappears from the leaves and the yellows to reds become evident.

The similar molecule mentioned above is called **xanthophyll** and has two oxygen atoms in addition to those of carotene. It is also present in autumn leaves and flowers and also reflects the longer wavelengths (absorbs the shorter). Carotene colors daffodils and dandelions, while xanthophyll colors marigolds and sunflowers.[19]

Another group of organic molecules, called **anthocyanins,** absorb in the middle of the visible spectrum and therefore give rise to many deep reds, blues, and purples in flowers. Most notably, roses derive their color from these, as does the cornflower.

One type of substance that we have not considered at all yet is a *metal.* We wish to consider metals briefly here because they do have special optical properties and these are qualitatively explained by our model of forced oscillators. The most obvious optical properties of metals are that they are very opaque and shiny when polished, even in very thin sheets such as aluminum foil, metals let through no light at all. The special optical properties of metals are due to free electrons within them, as are their special electrical and thermal properties. When many atoms of a metal are brought together to form a solid chunk, each one gives up one or more electrons from its outer shell, which are then free to move throughout the whole solid chunk and are called **free electrons.** Generally, there are still some electrons, called **bound electrons,** left attached to each atom. The relatively easy transport of the free electrons accounts for the high thermal and electrical conductivity of all metals. If we consider the free electrons as forced oscillators when a light wave strikes the metal, they must be thought of as being on a very, very weak spring, corresponding to the confinement only within the whole solid (much larger than an atom or even a large organic molecule). Therefore the applied frequency is much higher than the resonant frequency of the free electrons. According to our earlier discussion the electronic response should be small, but the important point is that it is *opposite* in phase to the incoming wave. If there is a high enough density of electric charges oscillating opposite in phase to the electromagnetic wave in a substance, they just cancel the wave; within about half a wavelength the net effect drops to zero. There is no refracted wave, but instead all the energy goes into the reflected wave, making the metal shiny.

A few metals, such as copper and gold, have a yellow or reddish color. This color is due to absorption, but not by the free electrons. Instead, in these cases the outer bound electrons have resonances in the blue-violet. These resonances do not correspond to jumps to the next Bohr orbit, which would be in the ultraviolet, but

[19] Sir William Bragg, *The Universe of Light* (New York: Dover Publications, Inc., 1959), p. 120.

rather to a rearrangement over smaller energy gaps within an incompletely filled orbit.

10.9 OTHER NATURAL LIGHTING EFFECTS

There are several other lighting effects in nature which are interesting in their own right but do not involve the production of color. It will be worth our while to take a little space here to consider them. The first is one we promised earlier to discuss because, like the glory, it leads to a halo around the shadow of the observer's head. This phenomenon is usually seen on dewey grass and is called the *heiligenschein* (German for *holy light*). When the sun is low in the morning and your shadow lies on dewy grass, often you can see this heiligenschein as an aureole or circular area of light on the grass around the shadow of your head. The sixteenth-century Italian artist Benvenuto Cellini thought it was a sign of his own genius.[20] When you look at the shadow of your head, you are looking in a direction exactly opposite to the sun, and therefore the reflected light has the least obscuring shadows. This fact, combined with the tendency of the dewdrops on the grass to act as retroreflectors (see Chapter 4), accounts for the enhanced brightness seen.[21] Figure 10–18 shows this latter effect.

Let the sun shine on a flat expanse of ground for a while and quite often an **inferior mirage** will be formed. This mirage was notorious in the desert for appearing as a large body of water glimmering in the distance. In modern times it is most often seen as a retreating pool of water on a paved road or highway. This familiar mirage appears when the air near the heated surface is hotter than that above; in some cases the change may be as much as 20°F to 30°F in the first half inch and then a few degrees per inch above that.[22] Hot air is less dense than cooler air and has a smaller refractive index, causing a bending of light as it comes down through the air. In order to see the direction of bending it is easiest to just picture one boundary between cooler

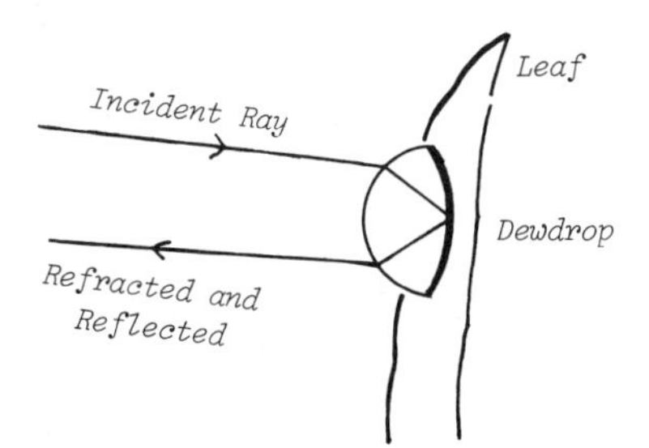

Figure 10–18 Dewdrops as Retroreflectors

[20] Minnaert, *Light & Colour in the Open Air*, p. 231.

[21] Ibid., p. 232.

[22] Ibid., p. 45.

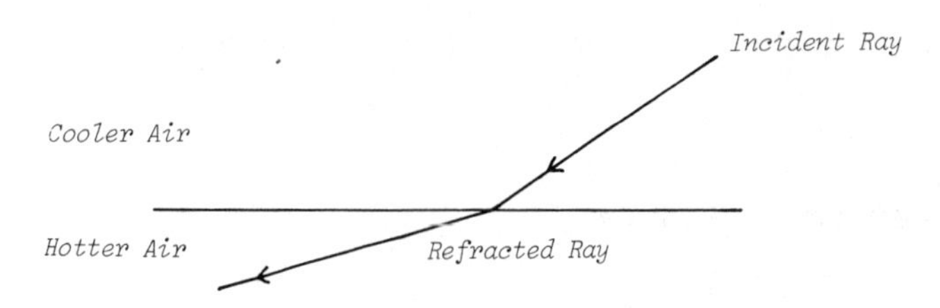

Figure 10–19 Refraction at cool air–hot air boundary

air above and hotter air below; the light bends away from the normal, as shown in Fig. 10–19. Of course in nature there is no such sharp boundary but rather a continuous change from cooler to hotter air and from higher to lower refractive index. Correspondingly, there is a continuous rather than sharp bending of the light rays, which can actually turn around near the hot surface and head upward, as shown in Fig. 10–20. What we actually see is not water reflections at all but refracted light from the sky. Furthermore the figure should make it clear that, as with the rainbow, when you approach the mirage you simply receive the same kind of light from farther away, so that the mirage appears to recede.

One other well-known natural effect, which has significance in art, is what is called the **moon illusion.** The rising or setting moon (or sun for that matter, but the moon is easier to look at) appears 2.5 to 3.5 times larger[23] than when it is high in the sky. This effect is definitely an illusion, since simple measurements show the angular diameter to be unchanged. There have been many attempted explanations of this illusion but none are completely satisfactory. It has been noted that the sky appears closer overhead than at the horizon, so the same size moon at the horizon is perceived as farther away and larger. Or, we are used to seeing objects in the sky such as birds move toward the horizon and get smaller as they move away. But the moon keeps the same size as it approaches the horizon and so appears to be actually getting larger. It has even been suggested that the attitude of the observer's head with respect to his or her body makes the difference: when you tilt your head back by bending your neck to

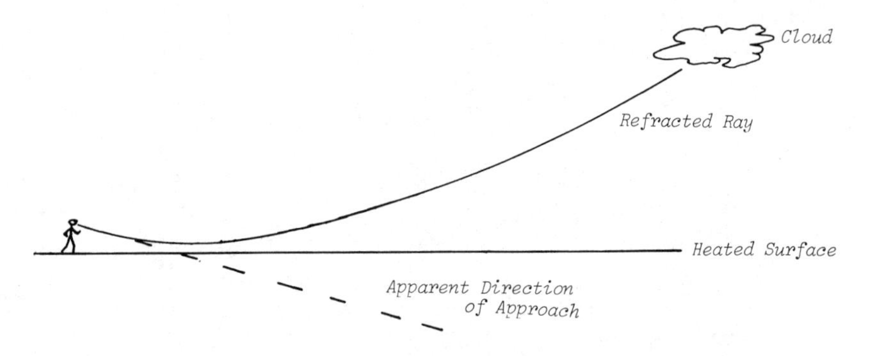

Figure 10–20 Formation of inferior mirage

[23] Ibid., p. 157.

view the moon, it appears smaller, but if you view it lying on your back with your neck straight, as you do when it is on the horizon, it appears larger.[24] Whatever the cause, the moon is almost always shown larger than reality in art. Indeed, if it were shown on the correct scale with respect to other objects in the picture, it would be nearly unnoticeable; it is no bigger than a quarter viewed from 115 cm away.

There are so many less common natural color and lighting effects that we could spend a good deal more time discussing them. But we must draw the line somewhere and get on to other things. The interested reader could do no better than to study Minnaert directly.

QUESTIONS

1. At what wavelength does the spectrum of sunlight peak?
2. If more mass is added to a weight holder on a spring, does its natural frequency increase, decrease, or remain the same?
3. If an oscillator is acted upon by a periodic force does it vibrate with the frequency of the force, its own natural frequency, or some intermediate frequency?
4. What applied frequency of a periodic force causes an oscillator to vibrate with the largest amplitude?
5. In which region of the spectrum is the usual resonant frequency of an electron vibrating in an atom? What about atoms vibrating in a molecule?
6. Why is the sky blue? What does the blue sky have to do with red sunsets?
7. If a bluish smog turns whiter as the day progresses, what can you say about the size of the smog particles?
8. Why does a snow accumulation look opaque when each crystal is clear?
9. What color are daytime shadows on snow? Why?
10. What causes the primary rainbow? Why are you unable to reach the "pot of gold" at the end of it?
11. What causes the secondary rainbow? Compare its appearance with that of the primary rainbow.
12. What are supernumerary rainbows?
13. What is dispersion?
14. Describe the skylight inside and outside of primary and secondary rainbows.
15. What is the 22° halo and what causes it?
16. What is the 46° halo and what causes it?
17. What are mock suns? What other names are given to them?
18. How can you tell the difference between a corona and the 22° halo around the moon?

[24] Ibid., p. 162.

19. What is the glory?
20. Give some examples of interference colors in nature.
21. Is chlorophyll green because it has a resonant frequency in the green part of the spectrum? Explain.
22. What is the connection between the colors of butter, egg yolks, dandelions, and carrots?
23. Why do leaves change from green to red and gold in the fall?
24. Why are roses red?
25. Why are metals very opaque?
26. What causes the color of copper and gold?
27. What is heiligenschein?
28. Describe an inferior mirage. What causes it?
29. What is the moon illusion?

Chapter 11

Color Science

11.1 NEWTON'S WORK

The scientific study of color in itself began with that great British genius of the seventeenth century, Sir Isaac Newton. Newton was the first to show that white light is composed of a mixture of all the colors of the rainbow by passing sunlight through a prism, as shown in Figure 10–5. More, however, was required than that simple demonstration to reach such a conclusion. Newton separated single colors from the rest and showed that further refraction in a prism did not alter them more, as may be seen from the experiment shown in Fig. 11–1. He showed further that recombining the colors with a second prism again produces white light, as illustrated by Fig. 11–2. Such experiments serve to show that the color is not being "added" to the light in any way by the prism (as in passing white light through a color filter), but rather is inherent in the white light from the start. Because they were not associated with any

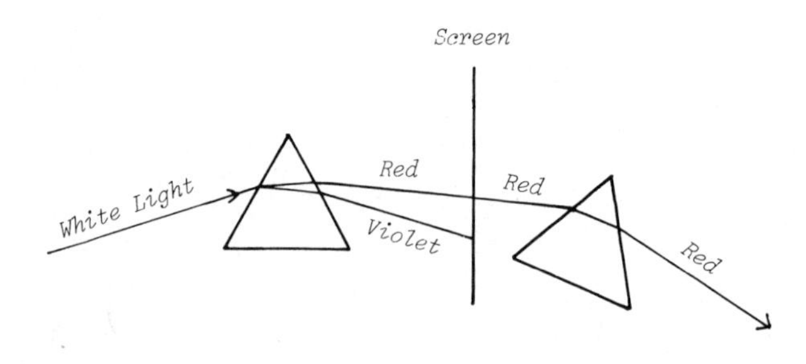

Figure 11–1 Second refraction of a single color

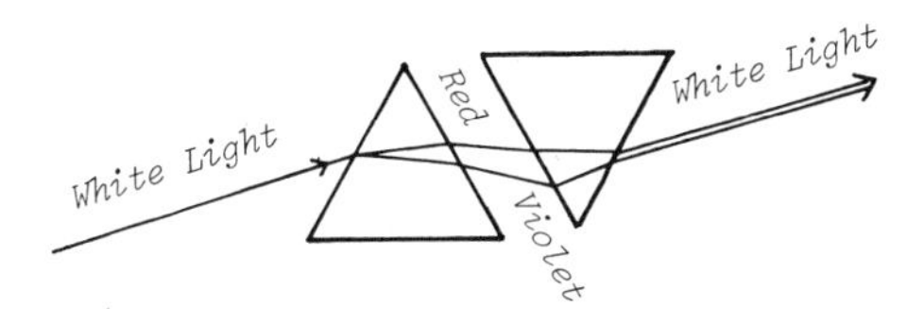

Figure 11–2 Recombination of the colors

physical object but seemed to be in the light itself, Newton called the colors **spectra** (specters or ghosts) and the collection of colors together a **spectrum.**[1]

Newton named seven colors in the spectrum: red, orange, yellow, green, blue, indigo, and violet. More commonly today we only speak of six major divisions, leaving out indigo. A careful reading of Newton's work indicates that the color he called indigo, we would normally call blue; his blue is then what we would name blue-green or cyan. Newton also noted that purples could be created by combining light from the two ends of the spectrum, red and violet. Since all the spectral colors merge rather imperceptibly one into the next and the purples can, in a similar manner, bridge the gap from red to violet, Newton introduced a color circle with no abrupt color changes, as shown in Fig. 11–3. White was represented at the center of the circle because it was a mixture of all colors; remember that we are still talking about mixing lights, not pigments. Now as colors become less "intense" or tended more toward white, they could be placed on a radius line to the correct hue but closer to the center. Rules for mixing colors could then be based on positions on or in the color circle. In particular, two colors on opposite sides of the circle could be mixed to produce white. This scheme was a good beginning and incorporates ideas we still use today, but much more work was required.

One reason that color science is a difficult field is that it combines the usually separate areas of physics and psychology. For example, in Newton's time the wave nature of light and its connection with color were not understood at all, and it took the general advance of physics to clear that up (see Chapter 2). But in the final analysis, when we speak of *color* we are speaking about human perceptions and sensations, clearly subjective and in the area of psychology.

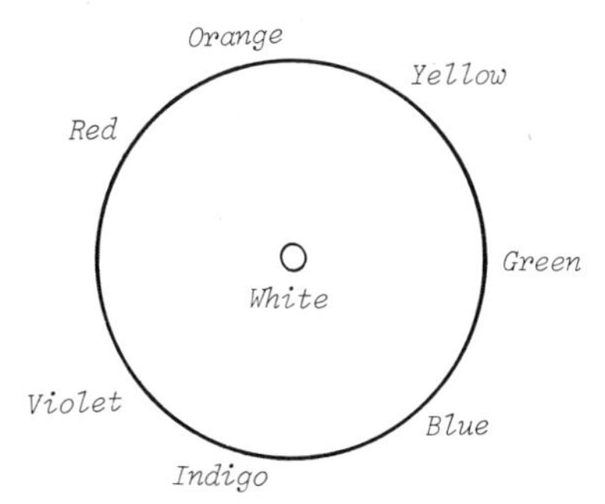

Figure 11–3 Newton's color circle

[1] David L. Mac Adam, *J. Opt. Soc. Am.* 65, (1975), 483.

11.2 PRIMARY COLORS

One very basic subjective aspect of color is that some seem to be **primary** or to come before the others. Those who have studied the development of language tell us that the names for black and white are invented first. From our point of view we can think of these two as the mixtures of no colors and all colors respectively; sometimes they are called **achromatic.** Next the color name for red appears, then yellow, green, and blue. Any other color names come much later in the evolution of the language. Notice that the colors that Newton chose to name in the spectrum include these four: red, yellow, green, blue. These four are almost always named also in any modern description of the spectrum or the rainbow. They seem to be subjectively distinct, whereas other colors of the spectrum seem to be mixtures. For example, cyan *subjectively appears* to be partly green and partly blue; orange has aspects of red and yellow, and so on.

A French contemporary of Newton, Edmé Mariotte (1620–1684) was one of the earliest to specifically state the basic premise upon which color science is now based: all colors can be produced by a mixture of just three primary colors. Mariotte chose as his primaries red, yellow, and blue. Thomas Young in his earliest descriptions of color vision also named red, yellow, and blue as primaries, but later substituted green for yellow. Actually there is a good deal of leeway as to which three colors are chosen as primary. The fundamental fact of color science is that almost all colors (and certainly all spectral colors) can be produced by a mixture of three primaries which satisfy the following conditions:[2] no one primary can be matched by a mixture of the other two, and all three in some proportion can produce white.

11.2.1 Additive Mixing

Mixing colored lights, as opposed to pigments, is called **additive mixing.** Usually the three primaries for additive mixing are chosen as red, green, and blue. Three such colored spotlights shown overlapping on a screen produce an appearance like that shown in Fig. 11-4. Here *R, G,* and *B* stand for red, green, and blue, respectively, while the overlap regions are labeled *C* for cyan (blue-green), *M* and magenta (purple), *Y* for yellow, and *W* for white. By adjusting the amounts of the primaries in the various mixtures, we can change the overlap colors so that any spectral color is matched. For example, the yellow region can be adjusted from a greenish yellow to an orange yellow by increasing either the proportion of green or that of red. In addition, **desaturated colors** (those Newton called less intense) can be produced in the white or triple overlap portion of the pattern by adjustment of one or more of the primaries. So all the colors of Newton's circle can be produced in this manner.

Two colors that are opposite to each other on Newton's color circle can be

[2]Ralph M. Evans, *The Perception of Color* (New York: McGraw-Hill Book Company, 1974), p. 44.

mixed to produce white and are said to be **complementary.** Yellow and blue (Newton's indigo), red and cyan, and green and magenta form complementary pairs. Note that those colors which are complementary to the primaries show up as the overlap colors in Fig. 11-4. If we represent white by the addition of two complementary colors, then one of the complementary pair can, in the same sense, be said to be white minus the other. Thus

$$W = Y + B$$

and $$Y = W - B$$

Also $$C = W - R$$

$$M = W - G$$

Often the complementaries to the primaries are simply said to be minus the primaries, with the white being understood.

$$Y = -B$$
$$C = -R$$
$$M = -G$$

This designation also makes sense from the point of view of Fig. 11-4. The white region there is where all three primaries are present. If we move from white to a region where one of the primaries is absent, we find ourselves in a color complementary to the missing primary.

11.2.2 Subtractive Mixing

The negative primaries (cyan, magenta, yellow) are the colors used for mixing of pigments, called **subtractive mixing.** This name is used because pigments work by *absorption,* taking out certain wavelengths from the incident light. When pigments are mixed, what adds up is the light *removed.* Thus a cyan pigment absorbs red and a yellow pigment absorbs blue; when cyan and yellow pigments are mixed the resulting pigment removes both red and blue from incident white light, reflecting green. Figure 11-5 shows subtractive mixing results.

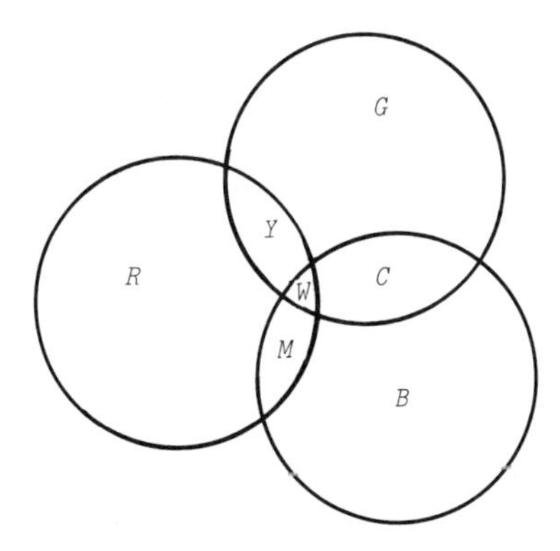

Figure 11-4 Additive primaries

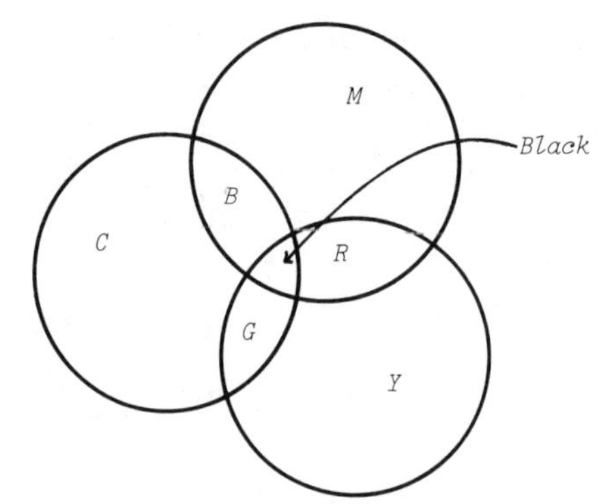

Figure 11-5 Subtractive primaries

The experimental facts of color science discussed above have very important implications. Consider the relationship of wavelengths to color. We can see that there is not a one-to-one correspondence: light with a wavelength of 580 nm always appears yellow, but yellow light (even the same shade of yellow) does not always consist of the one wavelength of 580 nm. A matching yellow might be produced by a combination of green and red light with *no* wavelengths of 580 nm. Such matching colors which differ in their spectrum (the amount of light at each wavelength) are called **metamers.** But the most practical result of the laws of color science is that they make color printing, color photography, and color television feasible. In each case we only need to work with three primaries instead of the vast array of colors which we wish to reproduce.

11.2.3 Examples of Primary Color Mixing

Color printing uses subtractive mixing of the negative primaries. Three separate printing plates (one for each negative primary) are used, which print the three inks in tiny dots of varying intensity. The eye integrates the dots to see different colors from different primary mixtures in different parts of the picture. Often a fourth plate in black ink is printed also to give shading and depth. Imagine trying to print in inks of all the different colors one wished to appear in the final picture!

Color photography was first demonstrated in 1861 by James Clerk Maxwell (see Chapter 2). Photography itself was only about 20 years old then. Maxwell took three different black and white photographs, each through a separate primary filter. When these were made into positive transparencies and projected, each through its own filter so the images overlapped on a screen, a full-color reproduction appeared. In this demonstration additive mixing from three different projectors was used to form the color image. Any area in the original scene that had a large proportion of red, say, would have sent a relatively large amount of light through the red filter and appeared light on that transparency. But that transparency was later projected through a red filter so that the corresponding area in the final image would receive a large amount of red light, as it should. Maxwell was limited by the fact that the black and white films then were far from equally sensitive to all colors, and this scheme was far too cumbersome for practical application anyway. However it did establish the possibility of color photography.

In modern **color photography** the filtering is all done within the film itself. There are three layers within the color film, sensitive to the three primary colors. Upon development, three dyes in the three negative primary colors are activated at the appropriate positions. The final color print is a subtractive mixture of these dyes.

Color television depends on additive mixing, the mixing of colored lights. There are actually three different phosphors in the television screen, arranged in tiny dots. These three phosphers produce red, green, or blue light when struck by the electron beam of the picture tube. As with the color print, the eye integrates the tiny dots of colored light to produce a mixture of primaries, so any color can be reproduced by stimulating the phosphors in proper proportion. Again, without the simplicity of only three primary colors color television would be next to impossible.

11.3 ATTRIBUTES OF COLOR

Before considering color mixing in a quantitative manner, it will be worth while to take a closer look at the naming and classification of colors. We have already seen how colors can be placed around and in a circle in a continuous manner, with white at the center. In that discussion we used the words **hue** and **saturation** without any careful definitions. In everyday language *hue* is often used as a synonym for *color,* but in color science we shall used *color* in a more general way. We may think of hue as the color variable to which we assign names, such as red, blue, yellow, and magenta. In other words, hue is what changes as you move around the color circle of Fig. 11-3 either on its outer circumference or on a smaller circle inside. All colors along one radius line of the color circle have the same hue. On the other hand, the colors along one radius line are perceptibly different, being closer to white, or less saturated, near the center, and further from white, or more saturated, at the circumference. Saturation is what varies as you move from the center of the circle outward. If either the hue or the saturation of two colors is different, we say the colors are different. With these definitions one can see that *hue* and *color* do not mean the same thing; all the colors along one radius of the color circle have the same hue but they are different colors because they have different saturations.

Hue and saturation are said to be **attributes** of color. But there is still (at least) one more attribute of color. That is, we can perceive two samples that seem to have the same hue and saturation but are still perceptibly different in color because they are different in another attribute. This third attribute of color is called **brightness** for colored lights or **lightness** for colored pigments reflecting light. The identification of brightness in the one case with lightness in the other is somewhat questionable,[3] but it will serve for our purposes as a first approximation.

The attribute of lightness might best be approached by considering a number of achromatic samples (perhaps paint chips) ranging from white to black through various shades of gray. Physically, each of these samples reflects all visible wavelengths equally, but different samples reflect different amounts: white reflects all wavelengths 100 percent, black 0 percent, and some gray 50 percent. These samples have zero saturation and no hue (hence the name achromatic). For example we have already placed white in our color circle at the center. Since saturation increases outward from the center in any direction, white at the center has zero saturation. Also, the center of a circle has no definable position around the circle, so we cannot assign any hue name to white.[4] In fact, the only attribute which distinguishes the samples is lightness. We could arrange them in order of increasing lightness from black to white.

11.4 COLOR SOLID AND COLOR ATLASES

These facts imply that the color circle is inadequate to represent all colors. Colors should be arranged in a three-dimensional space, corresponding to three attributes, instead of a two-dimensional space corresponding to two attributes. So we are led to a color solid to replace the color circle. The simplest way to obtain such a color solid is to imagine a number of color circles, similar to Fig. 11–3 and each corresponding to a different lightness, piled on top of one another. Now each color circle has, rather than white at its center, the achromatic color appropriate to the lightness of all colors in that circle. We still have hue varying around each circle and saturation varying outward from the center. Figure 11–6 shows such a color solid.

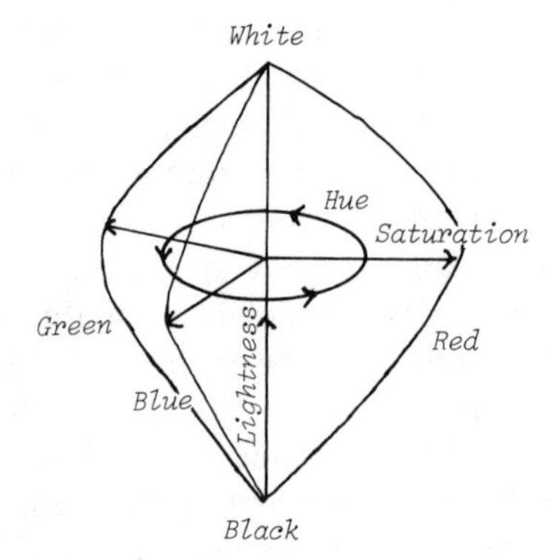

Figure 11–6 Color solid

[3] Ibid., p. 7.

[4] *The Science of Color* (Washington, D.C.: Optical Society of America, 1963), p. 66.

The three-dimensional nature of color is most graphically illustrated by various color atlases that have been published. These have colored samples arranged in an orderly way that demonstrates variation in hue, saturation, or lightness. One of the most famous and extensive is the *Munsell Book of Color*. This influential book has samples of 20 hues, each hue arranged with lightness varying vertically and saturation varying horizontally; each page would be similar to one vertical slice through the central axis of the color solid in Fig. 11–6, with a gray scale down the middle and complementary hues on each side of it. However, Munsell had his own naming system for colors: lightness is called **value** and saturation is called **chroma.** In this system hue is denoted by letters (for hue names) and numbers. The letters are *R, Y, G, B, P* for red, yellow, green, blue, and purple, respectively, and the combinations *YR, GY, BG, PB,* and *RP* are used for intermediate hues. But Munsell notation divides up the hue scale even more finely by using numbers from 1 to 10 within each of the 10 hue names, which leads to 100 hue divisions (not all reproduced in the atlas). The numbers are arranged so that each hue name falls on 5. Figure 11–7 illustrates the color circle with Munsell notation for hues. It may thus be seen that a hue specified as 10 *YR* is halfway between *YR* and *Y* or, what is the same thing, between 5 *YR* and 5 *Y*. Value and chroma are also specified by number, with value ranging from 0 to 10 (perfect white) and chroma ranging from 0 (achromatic) to some number specifying the most saturated pigment available at that hue and chroma. Thus a light, saturated greenish yellow might have the designation 4*GY* 6/10, where the 6 is the value and the 10 is the saturation. The *Munsell Book of Color* has some 960 samples in it.[5]

One problem with the Munsell system is that the notation is not very descriptive of the color, particularly to someone unfamiliar with the system. A system of standard color names has been developed by the Inter-Society Color Council (ISCC) and the U.S. National Bureau of Standards (NBS) and is illustrated by a small atlas of 251 color samples, the *ISCC-NBS Centroid Color Charts*. In this atlas the color samples are also arranged with a different hue on each page, lightness increasing upward and

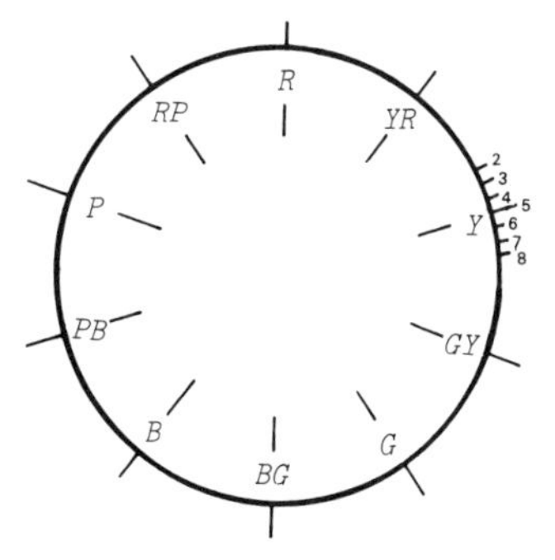

Figure 11–7 Munsell hues

[5] W. D. Wright, *The Measurement of Colour*, 4th ed. (New York: Van Nostrand Reinhold Company, 1969), p. 185.

saturation increasing away from a gray scale, but in this case the gray samples are on the left-hand side of the page and no complementary hue appears. In addition, each sample has a descriptive color name consisting of a hue name preceded by modifiers such as *dark, moderate,* or *light* to describe lightness, and *grayish, strong,* or *vivid* to describe saturation.

A more recent atlas produced by the Optical Society of America (OSA) concentrates on uniform steps in the three color attributes between samples. These *OSA Uniform Color Scales* contain about 500 such samples.

11.5 COLORIMETRY

In order to treat the measurement and specification of colors quantitatively it is now necessary to consider a scheme for assigning numbers to the amounts of primaries that make up a mixture. This quantitative treatment is known as *colorimetry* and has some subtly difficult features. Suppose we start with a color, *C,* which we wish to match in a suitable instrument by mixing appropriate amounts of the primaries red, green, and blue (additive mixing). The question immediately arises as to what units to use for measuring the amounts of red, green, and blue light. Our first thought might be to just use power units, that is, so many watts of red light, so many watts of green, and so on. But the information obtained in this way turns out to be not very useful.[6] Instead units for the three primaries are chosen so that equal amounts of the primaries mix to make a specified white light. In terms of color description this definition makes sense because, although in physical terms such as watts, one unit of red may be unequal to one unit of green, one unit of red mixed with one unit of green and one unit of blue always gives white. Using these units we might find R units of red mixed with G units of green and B units of blue matches our previously mentioned color C. Then R, G, and B are called the **tristimulus values** of C.

Usually we are only interested in the *proportion* of the primaries in the final mixture. For example we said white could be produced by one unit of each of the primaries mixed together. Another way of saying the same thing is to call white one-third red, one-third green, and one-third blue (or 0.333 red, 0.333 green, 0.333 blue). As another example, suppose our color C were matched by six units of red, three of green, and one of blue: it would be 0.6 red, 0.3 green, and 0.1 blue. We can summarize these results in the equations

$$(C) = r\cdot(R) + g\cdot(G) + b\cdot(B) \tag{16}$$

$$\text{with } r = \frac{R}{R+G+B} \tag{17a}$$

[6] Ibid., p. 82.

$$g = \frac{G}{R + G + B} \tag{17b}$$

$$b = \frac{B}{R + G + B} \tag{17c}$$

$$\text{and } r + g + b = 1 \tag{18}$$

Here r, g, and b stand for the fractions of red, green, and blue that go into the mixture for the color C. In Eq. (16), the colors themselves are indicated by letters with parentheses around them and the equation just expresses the color match between C and the mixture that is fractionally r red, g green, and b blue. The fractions r, g, and b are called the **chromaticity coordinates** of C. Equation (17) expresses the fact that the chromaticity coordinates can be easily found from the tristimulus values of C. Finally, Eq. (18) says that as the fractions that make a whole must add up to unity, so must the chromaticity coordinates of a color. It is important to remember that in these equations r, g, b, R, G, and B are numbers, but (R), (G), (B), and (C) stand for colored lights.

Example:

A color is matched by a mixture of two units of red light, one unit of green light, and five units of blue light. What are its tristimulus values and chromaticity coordinates in those primaries?

The tristimulus values are what is directly given:

$$R = 2, G = 1, B = 5$$

From Eq. (17)

$$r = \frac{2}{8} = 0.25$$

$$g = \frac{1}{8} = 0.125$$

$$b = \frac{5}{8} = 0.625$$

When a color is expressed in terms of its chromaticity coordinates, Eq. (18) implies that two coordinates will suffice; the other can be found by subtracting from 1. If a color is 0.6 red and 0.1 blue, it must necessarily be 0.3 green.

The next order of business in colorimetry is to find the chromaticity coordinates of the spectral colors. To do that three primary colors must first be chosen; as noted earlier there is a good deal of leeway in this choice, and the results for one set of primaries can be manipulated mathematically to provide data for other sets of primaries. Then a number of different observers must match the spectral colors (single wavelength lights) with mixtures of the three primaries, and the results must be averaged. This task was carried out by a number of investigators in the 1920s. One interesting result is that no matter what set of primaries is chosen, not all spectral colors can be exactly matched. Instead, the primary mixtures are often somewhat desaturated compared with the spectral color, particularly in the blue-green; thus mixing the blue and green primaries might give the same hue as a spectral blue-green ($\lambda \cong 480 - 520$ nm) but lower saturation. However it was found that a small amount of the red primary could be added to the stimulus (the spectral color), desaturating it enough to be matched by a mixture of the blue and green primaries. In this case, we say a *negative* amount of red is in the mixture. In other words we allow negative tristimulus values and chromaticity coordinates. With this provision all spectral colors can be matched, and Eqs. (16) through (18) still hold.

The results of the investigations described above are often displayed on a graph of two of the chromaticity coordinates for the spectral colors, which is called a **chromaticity chart** or diagram. Remember that the third chromaticity coordinate, say b, can always be found from the two plotted in the chromaticity chart by

$$b = 1 - r - g \tag{19}$$

Figure 11–8 shows such a chromaticity chart for primaries chosen as the spectral colors with wavelengths 650 nm (red), 530 nm (green), and 460 nm (blue). In this diagram the curved line shows the location of the chromaticity coordinates of the indicated spectral colors. Note that 650 nm lies right at 1 on the red primary axis, 530 nm lies right at 1 on the green primary axis, and 460 nm lies right at (0,0), the origin. These points must be in these positions because we know that spectral light at 650 nm should be matched by the red primary with zero proportion of green and blue ($r = 1$, $g = 0$, $b = 0$); similar arguments hold for green and blue, in view of Eq. (19). Also the negative red chromaticity coordinates can be clearly seen on the left-hand side of the figure.

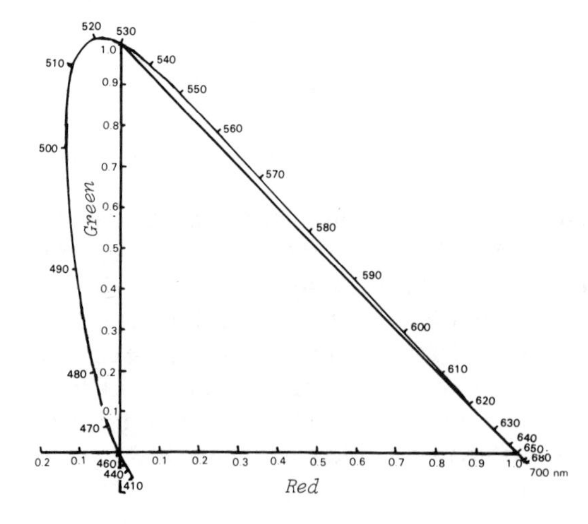

Figure 11–8 Chromaticity chart for primary wavelengths 650, 530, and 460 nm

Example:

With the primaries of Fig. 11-8, what are the chromaticity coordinates of light of wavelength 560 nm?

Find the point marked 560 on the curve in the figure. Read straight down on the red axis for $r = 0.25$. Read straight across on the green axis to find $g = 0.78$. Now Eq. (19) says

$$b = 1 - g - r = 1 - 0.78 - 0.25$$
$$b = -0.03$$

11.5.1 CIE System

Since light of a single wavelength gives the most saturated color possible, the curve in the chromaticity diagram marks the boundary of possible saturation. Less saturated colors should be placed inside that boundary toward white, which is centrally placed. We have a diagram similar to Newton's color circle, but it has become horseshoe-shaped to accommodate the quantitative color mixing data gathered over the years. The mixture (additive) of any two colors lies on a line joining the two in the chromaticity chart. In Fig. 11-8 the straight lines forming the right triangle with the axes as two sides show the boundaries for colors that can be matched by *positive* mixtures of those primaries. The triangle lies totally inside the horseshoe spectral curve, which indicates that there are some colors (inside the horseshoe curve but outside the triangle) that cannot be produced by positive mixtures of these primaries, which we already knew. Furthermore any other three real primaries, since they must lie in or on the horseshoe curve, will also form mixtures that are limited by a triangle totally inside the horseshoe. Any such primaries will lead to negative tristimulus values and chromaticity coordinates for some physically realizable colors. But what if we chose primaries outside the horseshoe curve? We should make clear from the start that any such primaries will not be physically realizable colors, all of which lie in or on the horseshoe. Colors outside the curve represent those further from white, and therefore more saturated, than single wavelength light, which is impossible—they are a mathematical trick. However, if we could work with such nonphysical colors, we could pick three that form a triangle in the chromaticity chart that completely encloses the horseshoe curve of physical colors. Then all physical colors would have *positive* tristimulus values and chromaticity coordinates in terms of those primaries, not determined directly from experiment but calculated from experimental data with real primaries. The problem of desaturation when primaries are mixed is solved by starting with primaries that are more saturated than any physical colors to be matched.

In 1931 the Commission Internationale de l'Éclairage (CIE) took just such steps to define a standardized color coordinate system with no negative values. The chromaticity chart for primaries with wavelengths of 700 nm, 546.1 nm, and 435.8

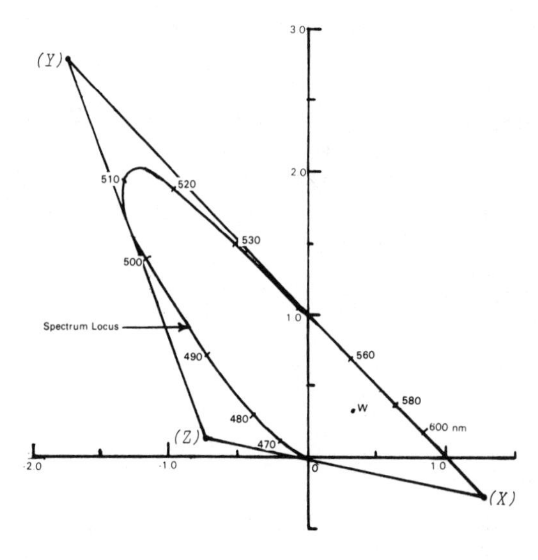

Figure 11-9 Chromaticity chart for primary wavelengths 700, 546.1, and 435.8 nm with CIE standard primaries (*XYZ*)

nm is shown in Fig. 11-9 along with the nonphysical primaries chosen, which are labeled (*X*), (*Y*), and (*Z*).

If we are going to specify colors in the *XYZ* system, there is little point in having the *R* and *G* axes present. A mathematical transformation (beyond the scope of this text) takes us from the *R* and *G* axes to the X and Y axes, now shown at right angles in Fig. 11-10, with the horseshoe curve of physical colors also shown. The transformation which takes us from Fig. 11-9 to Fig. 11-10 is equivalent to looking at Fig. 11-9 tilted, so that the X and Y axes appear to be at right angles. If you try that you will see the horseshoe shape of Fig. 11-9 become that of Fig. 11-10.

Example:

What are the CIE standard chromaticity coordinates of 560 nm wavelength light?

Find the point marked 560 nm on the curve in Fig. 11-10. Read straight down to find $x = 0.37$ and straight across to find $y = 0.62$.

$$\text{Now } z = 1 - x - y = 1 - 0.37 - 0.62$$
$$z = 0.01$$

You should never get negative values from this chart!

With the aid of the CIE standard chromaticity diagram colors are often specified in terms of **dominant wavelength** and **purity** instead of hue and saturation. These concepts allow a quantitative designation of hue and saturation. Suppose we have a color *C* whose dominant wavelength and purity we wish to find. First it must be correctly placed in the chromaticity diagram. Next a line must be drawn from

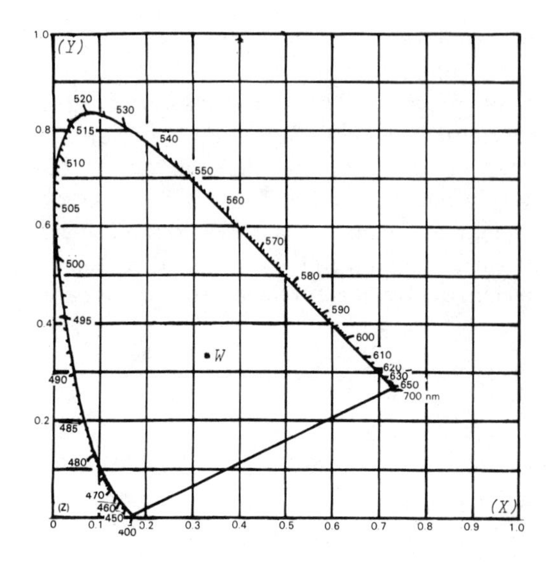

Figure 11–10 CIE standard chromaticity chart

white W through C to the spectral color curve, as shown in Fig. 11–11. The point at which the line intersects the horseshoe curve gives the dominant wavelength, in our illustrated case, 540 nm. This result tells us that the particular color C is somewhat yellowish green in hue. The purity is found from the ratio of lengths along the line, WC divided by the total length of the line from W to 540 nm on the curve. We can use a simple ruler to find those lengths in our case, giving a purity value p.

$$p = \frac{8 \text{ mm}}{26 \text{ mm}}$$

$$p = 0.308$$

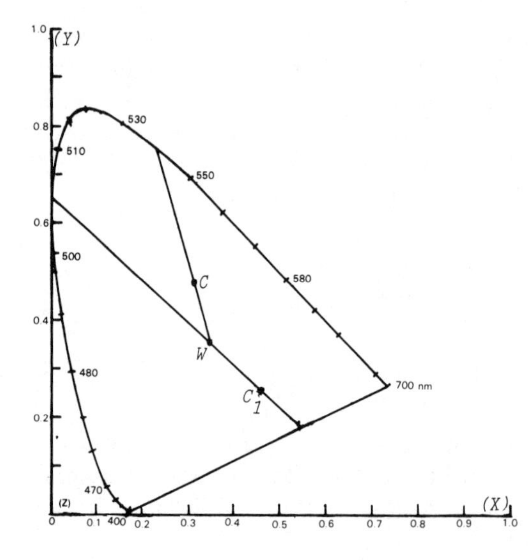

Figure 11–11 Finding dominant wavelength and purity

Note that the purity can range from 0 (if C is right on W) up to 1 (if C is right on the spectral curve). Thus small values of p represent unsaturated colors and values near 1 represent saturated colors. In our example, C is a rather unsaturated greenish yellow.

A slightly different procedure is required if the color in question is purplish so that the line from W does not intersect the horseshoe curve itself but rather the straight line joining the spectrum ends. The point C_1 in Fig. 11–11 represents such a color. Now there is no wavelength at the point of intersection; no single wavelength gives purple. Instead we just extend the line in the opposite direction as shown and read the complementary wavelength. The measurement for purity does not change.

QUESTIONS

1. Who first showed white light is a mixture of all the colors of the rainbow?
2. What four colors seem to be primary, or unmixed in sensation?
3. What is additive mixing of colors? What are the primaries for additive mixing?
4. What color is complementary to magenta?
5. What is subtractive mixing of colors? What are the primaries for subtractive mixing?
6. What are metamers?
7. Who first demonstrated the possibility of color photography?
8. What kind of color mixing does color television use?
9. What are the three attributes of color?
10. What colors have zero saturation?
11. What are the Munsell names for lightness and saturation?
12. Describe the color given by the Munsell notation 5Y 8/12.
13. If a color is matched by a mixture of three units of red light, seven of green light, and two of blue light, what are its tristimulus values and chromaticity coordinates in terms of those primaries?
14. In terms of the primaries of Fig. 11–8, what are the chromaticity coordinates of light of the single wavelength 500 nm? Which of the primaries must be used to desaturate the sample light to obtain a match?
15. What are the CIE standard chromaticity coordinates of light of the single wavelength 500 nm?
16. In what color attribute do the CIE standard primaries surpass all physically realizable light?
17. Find the complementary dominant wavelength and purity of the color C_1 in Fig. 11–11.
18. Use Fig. 11–10 to find the dominant wavelength and purity of a color with CIE standard chromaticity coordinates $x = 0.55$, $y = 0.39$ (assume white is at $x = 0.333$, $y = 0.333$ as shown). Describe the color in words.

Appendix A:

Lens and Mirror Equations

One equation relates the object distance p, the image distance q, and the focal length f for both thin spherical lenses and spherical mirrors. It states,

$$\frac{1}{p} + \frac{1}{q} = \frac{1}{f} \tag{1}$$

Before using this equation, there are two points to be aware of. Any length units will work here, as long as the same units are used for p, q, and f. Most importantly, Eq. (1) is *not* at all the same thing as saying $p + q = f$, any more than writing $1/4 + 1/4 = 1/2$ (correct) is the same thing as saying $4 + 4 = 2$ (incorrect). To solve for any one of the three quantities in Eq. (1) we must perform the appropriate mathematical operations with fractions.

In most practical problems p and f are given and q must be found. Then we must solve Eq. (1) for q in terms of the other two quantities. First we subtract $1/p$ from both sides to get $1/q$ alone.

$$\frac{1}{q} = \frac{1}{f} - \frac{1}{p}$$

Next the two fractions on the right must be put over a common denominator pf.

$$\frac{1}{q} = \frac{p}{pf} - \frac{f}{pf}$$

$$\frac{1}{q} = \frac{(p - f)}{pf}$$

Now, since both sides are in the form of a single fraction, we may invert them to get our solution for q:

$$q = \frac{pf}{(p - f)} \quad (2)$$

Equation (1) is said to be **symmetric** in the variables p and q. That means that the symbols can be interchanged and the mathematical sense of the equation remains unchanged: it still says the same thing. This mathematical invariance simply reflects the physical fact that the object and image are interchangeable, which stems from the still more basic reversibility of light rays, as discussed in Chapter 4. We may use this symmetry in p and q to transform Eq. (2) into a solution for p by interchanging p and q.

$$p = \frac{qf}{(q - f)} \quad (3)$$

These equations are valid for all cases, with the distances taken as positive for real objects, images, and foci and negative for virtual images and foci (see Chapter 4).

Example:

Find the image position when an object is placed 20 cm away from a thin, spherical converging lens with focal length 15 cm.

Equation (2) is already solved for q

$$q = 20 \times \frac{15}{(20 - 15)} = 20 \times \frac{15}{5}$$

$$q = 20 \times 3 = 60 \text{ cm}$$

The positive result for q implies a real image on the opposite side of the lens from the object.

Example:

Find the image position of an object placed 12 cm away from a concave mirror with a 36-cm focal length.

Again using Eq. (2)

$$q = 12 \times \frac{36}{(12 - 36)} = \frac{12 \times 36}{(-24)}$$

$$q = \frac{36}{(-2)} = -18 \text{ cm}$$

This must be a virtual image behind the mirror because of the negative sign.

Example:

Where is the image of an object placed 20 cm from a diverging lens with a focal length of 15 cm?

A diverging lens, like a convex mirror, has a virtual focus, so f must be made a negative number in Eq. (2):

$$q = 20 \times \frac{-15}{20 - (-15)}$$

$$= -20 \times \frac{15}{(20 + 15)}$$

$$= \frac{-300}{35}$$

$$q = -8.57 \text{ cm}$$

Another virtual image.

There are other equations that are often used in conjunction with these. For instance the image size (linear dimension) can be found simply from:

$$\frac{I}{O} = \left|\frac{q}{p}\right| \tag{4}$$

The straight lines indicate absolute value: take q and p as positive always. I is image size and O is object size.

Example:

If the object in the mirror example above is 3 cm high, how large is the image?

From Eq. (4):

$$I = O \left|\frac{q}{p}\right|$$

$$I = 3 \times \frac{18}{12} = 4.5 \text{ cm high}$$

Equations of about the same level of difficulty as Eq. (1) also allow us to find focal lengths from more basic information about spherical mirrors or thin lenses. For mirrors, the equation is particularly simple

$$\frac{1}{f} = \frac{2}{R} \tag{5a}$$

$$\text{or} \quad f = \frac{R}{2} \tag{5b}$$

where R is the radius of curvature of the mirror surface. Sometimes Eq. (5a) is used because $1/f$ enters into Eq. (1) and because $1/f$ is an easier form to remember for lenses.

The corresponding expression for the focal length of a thin lens (in air) is more complicated for two reasons: the law of refraction is more complicated than the law of reflection, involving the index of refraction, and the light rays change direction at two surfaces rather than just at one. The lens equation is

$$\frac{1}{f} = (n - 1)\left(\frac{1}{R_1} + \frac{1}{R_2}\right) \tag{6}$$

where n is the refractive index of the lens material, and R_1 and R_2 are the radii of curvature of the two lens faces.

Both Eq. (5) and Eq. (6) must always be used with a sign convention in order to end up with the correct sign for f (real or virtual focus). Unfortunately the sign convention for mirrors is opposite to that for lenses. For mirrors, concave faces are always assigned a positive radius of curvature and convex faces a negative one. For lenses, concave faces are always assigned a negative radius of curvature and convex faces a positive one. Note that the two values R_1 and R_2 for a lens may have different signs.

Example:

A convex spherical mirror has a radius of curvature of 64 cm. What is the mirror's focal length?

Convex for mirrors means a negative sign for R:

$$f = \frac{R}{2} = \frac{-64}{2}$$

$$f = -32 \text{ cm}$$

The minus sign indicates the virtual focus we know a convex mirror to have.

Example:

A lens is made from glass with $n = 1.5$ and has a convex face with radius of curvature 20 cm and a concave face with radius of curvature 30 cm. What is the focal length?

Here the sign convention for lenses implies $R_1 = +20$ cm and $R_2 = -30$ cm (which you call R_1 and which R_2 will make no difference in the answer).

$$\frac{1}{f} = (1.5 - 1)\left(\frac{1}{20} - \frac{1}{30}\right)$$

$$= 0.5\left(\frac{3}{60} - \frac{2}{60}\right)$$

$$= \frac{0.5}{60}$$

$$f = \frac{60}{0.5} = 120 \text{ cm}$$

This is a positive focal length, indicating a real focus and a converging lens.

Equations (5) and (6) will also work for plane surfaces if we take the reasonable position that a plane is the same thing as the surface of a sphere with an infinite radius of curvature ($R = \infty$). If we do not worry too much about mathematical rigor, we can then say $1/R = 0$. This approach says that for a plane mirror we have

$$\frac{1}{f} = \frac{2}{R} = 0$$

$$\frac{1}{p} + \frac{1}{q} = \frac{1}{f} = 0 \qquad (7)$$

$$\frac{1}{q} = \frac{-1}{p} \tag{8}$$

Then we invert both sides of Eq. (8) to get

$$q = -p \tag{9}$$

The image distance equals the object distance but is negative to indicate a virtual image, a result which should already be familiar.

Appendix B:

Snell's Law

The law governing refraction of light rays is called **Snell's law** after Willebrod Snellius (1591–1626), who first discovered it long before light was known to have wave properties. Snell's law involves a trigonometric function of the angles of incidence and refraction; therefore a brief introduction to that part of trigonometry is included here.

Consider the right triangle shown in Fig. B–1. The sine of the angle θ is defined as the ratio of two sides of the triangle:

$$\sin\theta = \frac{a}{c} \tag{1}$$

Now you might think that this definition does not make much sense because you could have drawn a different size triangle with the same angle θ but different length sides a and c. However, one reason this trigonometric function, $\sin\theta$, is important is precisely because, no matter what size you make the triangle, as long as θ says the

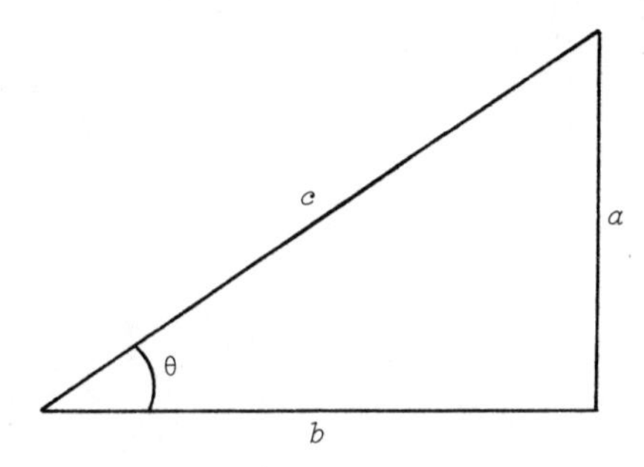

Figure B–1 Right triangle

same, the *ratio a/c* stays the same; that is, if you increase a by some factor, and then close the right triangle, c will be increased by the same factor. So $\sin\theta$ really does just depend upon θ, and values for $\sin\theta$ can be tabulated versus θ or read from a calculator that has trigonometric functions. Note that the sine of any angle is always less than or equal to 1. Table B–1 is a brief table of $\sin\theta$ versus θ.

Although there are other trigonometric functions, $\sin\theta$ is the only one you need to know for Snell's law. To avoid confusion over assigning symbols to the sides of the triangle some people remember the sine of θ as the side opposite to θ over the hypotenuse (longest side, or the one opposite the right angle).

In Fig. 4–3 let the refractive index of medium 1 be n_1 and that of medium 2 be n_2. Then Snell's law may be written in the form

$$n_1 \sin\theta_1 = n_2 \sin\theta_2 \tag{2}$$

When the law is formulated in this manner, there is no need to worry about which angle is the angle of incidence and which the angle of refraction; since light rays are reversible, the arrows could be turned around anyway.

If medium 1 is air, we know $n_1 \cong 1$, so Eq. (2) becomes

$$\sin\theta_1 = n_2 \sin\theta_2 \tag{3}$$

$$\text{or} \quad \sin\theta_2 = \frac{\sin\ \theta_1}{n_1} \tag{4}$$

TABLE B–1 SINE VALUES

Angle	Sine
0°	0
5	0.087
10	0.174
15	0.259
20	0.342
25	0.423
30	0.5
35	0.574
40	0.643
45	0.707
50	0.766
55	0.819
60	0.866
65	0.906
70	0.940
75	0.966
80	0.984
85	0.996
90	1.0

These equations can be used to find one of the angles if both refractive indices and the other angle are known.

Example:

Suppose that in Fig. 4–3, medium 1 is air and medium 2 is water. Find θ_2 if $\theta_1 = 30°$.

Equation (4) and Table 4–1 are used to find

$$\sin\theta_2 = \frac{\sin 30°}{1.33}$$

$$= \frac{0.5}{1.33} \text{ (from Table B–1)}$$

$$= 0.376$$

Using Table B–1 in reverse gives

$$\theta_2 \cong 22°$$

Example:

A light ray in going from water into quartz makes an angle of incidence of 40°. Find the angle of refraction.

We may assign medium 1 to be water and medium 2 to be quartz. Table 4–1 tells us that $n_1 = 1.33$ and $n_2 = 1.54$. Now θ_2 is the angle of refraction.

$$\sin\theta_2 = \left(\frac{n_1}{n_2}\right)\sin\theta_1$$

$$= \left(\frac{1.33}{1.54}\right)\sin 40°$$

$$= \left(\frac{1.33}{1.54}\right) \times 0.643 \text{ (Table B–1)}$$

$$= 0.555$$

Again from Table B–1:

$$\theta_2 \cong 34°$$

Appendix C:

Scientific Notation

Often in this course (and in many other science courses) we have occasion to use very large or very small numbers. An example of a very large number would be the speed of light expressed in meters per second (m/s): $c = 300{,}000{,}000$ m/s. An example of a very small number might be the wavelength of violet light in meters: $\lambda = 0.0000004$ m. Indeed, we may come across numbers very much larger or smaller than these examples. One can see that writing them in the ordinary decimal form as we have done wastes valuable time and space. **Scientific,** or powers of 10, **notation** is a condensed way of writing such numbers, a kind of shorthand.

The method is based on the powers of 10. Most students are familiar with the positive powers: $10^1 = 10$, $10^2 = 100$, $10^3 = 1000$, $10^4 = 10{,}000$, etc. There is obviously no limit to how high we could go with this sequence. You can see right away that very large numbers might be written as powers of 10 with greater economy than in the decimal form.

Negative powers are less familiar to most students. A negative power means 1 divided by the same positive power: $10^{-1} = 1/10 = 0.1$, $10^{-2} = 1/100 = 0.01$, $10^{-3} = 0.001$, $10^{-4} = 0.0001$, etc. In this case there is no limit to how small this sequence could become. Again there is a considerable savings in time and space by writing very small numbers in the powers of 10 form over using the decimal form.

Although it may not be obvious, this way of defining negative powers is just what is required to make our algebraic rules for multiplying and dividing powers of numbers always work correctly. For the same reason, any number raised to the zero power is defined as unity. More particularly in our case, $10^0 = 1$. We shall return to the multiplication and division rules shortly.

There is a quick way to convert back and forth between powers of 10 and the decimal form; there is no need to multiply together many factors of 10 in the process.

For the positive powers, the **exponent** (the power to which 10 is raised) tells you the number of zeros following the 1. For example, if you wish to convert 1,000,000 to powers of 10, just count the zeroes and then write 10^6. For the negative powers the exponent tells the decimal place in which the 1 appears. Thus 10^{-5} can be immediately converted to 0.00001 and vice versa. In other words, in order to convert between powers of 10 and decimal form you do not even have to be able to multiply and divide: all you have to know is how to count! That is not so surprising when you realize that scientific notation is merely a trick for keeping track of the decimal point, and that is basically a counting process.

Up to this point we have converted only numbers consisting of 1s and 0s. The reader may still wonder how we deal with numbers such as the speed of light and wavelength examples given earlier. Consider the speed of light: we may write 300,000,000 as $3 \times 100{,}000{,}000$. Now by our counting of zeros, $3 \times 100{,}000{,}000 = 3 \times 10^8$. This is the form in which the number is written in powers of 10 notation. It may be confusing to students who have never used it before, because the multiplication sign seems to indicate that there is still an operation to be performed. Despite that, 3×10^8 is really the final form and simply means count over eight places to the right of the 3 to find the correct position for the decimal point. Similarly, $0.0000004 = 4 \times 0.0000001 = 4 \times 10^{-7}$.

When numbers in scientific notation must be multiplied or divided, two steps must be remembered. First, perform the indicated operations on the numbers in front of the powers of 10 (e.g., the 3 in 3×10^8), ignoring the powers of 10, to arrive at a new number in the answer. Second, perform the operations indicated on the powers of 10 independently to arrive at a new power of 10 in the answer. In regard to the second step, it must be remembered that to multiply or divide powers of any number (in our case the number is 10), it is only necessary to add or subtract exponents, respectively.

Example:

What is the speed of light divided by the wavelength of violet light?

We wish to calculate $(3 \times 10^8 \text{ m/s})/(4 \times 10^{-7} \text{ m})$.

First divide 3 by 4: $\frac{3}{4} = 0.75$

Then calculate $\frac{10^8}{10^{-7}} = 10^{8-(-7)} = 10^{8+7} = 10^{15}$

The answer is 0.75×10^{15}/s

Often the decimal point in the first number is moved so that the number is between 1 and 10. To do that here we must multiply 0.75 by 10; in order to have the same answer 10^{15} will have to be divided by 10.

$$0.75 \times 10^{15}/\text{s} = 7.5 \times 10^{14}/\text{s}$$

Example:

Multiply 6.62×10^{-34} by 7.5×10^{14}

First $6.62 \times 7.5 = 49.7$

Then $10^{-34} \times 10^{14} = 10^{-34+14} = 10^{20}$

The answer is 49.7×10^{20}

Reduce the first number by a factor of 10 and increase the second by the same factor.

$$49.7 \times 10^{20} = 4.97 \times 10^{21}$$

Index

A

B

C

D

E

F

G

H

I

J

K

L

M

N

O

P

Q

R

S